Competition Law, Innovation and Antitrust

NEW HORIZONS IN COMPETITION LAW AND ECONOMICS

Series Editors: Steven D. Anderman, *Department of Law, University of Essex, UK* and Rudolph J.R. Peritz, *New York Law School, USA*

This series has been created to provide research based analysis and discussion of the appropriate role for economic thinking in the formulation of competition law and policy. The books in the series will move beyond studies of the traditional role of economics – that of helping to define markets and assess market power – to explore the extent to which economic thinking can play a role in the formulation of legal norms, such as abuse of a dominant position, restriction of competition and substantial impediments to or lessening of competition. This in many ways is the *new horizon* of competition law policy.

US antitrust policy, influenced in its formative years by the Chicago School, has already experienced an expansion of the role of economic thinking in its competition rules. Now the EU is committed to a greater role for economic thinking in its Block Exemption Regulations and Modernisation package as well as possibly in its reform of Article 82. Yet these developments still raise the issue of the *extent* to which economics should be adopted in defining the public interest in competition policy and what role economists should play in legal argument. The series will provide a forum for research perspectives that are critical of an unduly-expanded role for economics as well as those that support its greater use.

Titles in the series include:

Antitrust, Patents and Copyright
EU and US Perspectives
Edited by François Lévêque and Howard Shelanski

Innovation Markets and Competition Analysis
EU Competition Law and US Antitrust Law
Marcus Glader

Competition Law and Patents
A Follow-on Innovation Perspective in the Biopharmaceutical Industry
Irina Haracoglou

Antitrust and Regulation in the EU and US
Legal and Economic Perspectives
Edited by François Lévêque and Howard Shelanski

Competition Law, Innovation and Antitrust
An Analysis of Tying and Technological Integration
Hedvig Schmidt

Competition Law, Innovation and Antitrust

An Analysis of Tying and Technological Integration

Hedvig Schmidt

University of Southampton, UK

NEW HORIZONS IN COMPETITION LAW AND ECONOMICS

Edward Elgar

Cheltenham, UK • Northampton, MA, USA

Published by
Edward Elgar Publishing Limited
The Lypiatts
15 Lansdown Road
Cheltenham
Glos GL50 2JA
UK

Edward Elgar Publishing, Inc.
William Pratt House
9 Dewey Court
Northampton
Massachusetts 01060
USA

A catalogue record for this book is available from the British Library

Library of Congress Control Number: 2009936232

Mixed Sources
Product group from well-managed
forests and other controlled sources
www.fsc.org Cert no. SA-COC-1565
© 1996 Forest Stewardship Council

ISBN 978 1 84844 632 8

Typeset by Cambrian Typesetters, Camberley, Surrey
Printed and bound by MPG Books Group, UK

Contents

Figures and tables

FIGURES

TABLES

Acknowledgements

This book is based on my doctoral thesis, defended in March 2008, although updates and changes have been made. As a result, the scope and focus of the book are slightly different from those of the thesis. In the process of writing both thesis and book, I have received great support and encouragement from a number of people and institutions. In particular, I would like to express my heartfelt gratitude to Professor Steve Anderman (University of Essex), my thesis supervisor, who provided encouragement, invaluable guidance and fruitful discussions over the years. He is also the person to whom I undoubtedly owe my academic career.

My thanks also go to Giorgio Monti (London School of Economics) and Professor Rob Merkin (University of Southampton) for their support and helpful comments on the draft of the book at various stages.

I am grateful to the Max Planck Institute in Munich, Germany, and its employees, who gave me a five month research scholarship at the institute – they were probably the best months of my time as a PhD student. This was not least thanks to Dr Christopher Heath (European Patent Office), Professor Stefan Enchelmaier (York Law School) and Dr Beatriz Conde Gallego (MPI). Furthermore a special thank you goes to the Law Department, University of Essex, Nordea Danmark Fonden and Den Schou Beckmanske Stifltelse which financially supported my research.

On a more personal note, I would like to thank my 'old PhD friends' Estelle Askew-Renault and Beate Lichtwardt, as well as my colleagues at the University of Southampton School of Law: Ed Bates, Oren Ben-Dor, Gerrit Betlem, Nick Hopkins, Emma Laurie, Renato Nazzini, Emily Reid, Mark Telford and Caroline Wilson for their helpful comments, discussions, and not least for keeping me sane throughout the final years of writing.

A special thank you goes to Tina Christensen, Christina Hall Frederiksen and Kim Vejen Larsen, without whose wisdom, encouragement and whip I would never have embarked upon a PhD – so: thank you, it meant the world!

Finally, to my family: Paul, my (patient and ever-supportive) other half, my parents, Gunver and Ole, and, last but not least, my sister Anne: thank you for being there for me during the ups and downs.

Abbreviations

CFI	Court of First Instance
CMLR	Common Market Law Reports
DoJ	US Department of Justice
EC	European Community
ECJ	European Court of Justice
ECR	European Court Reports
EC Courts	Court of First Instance and European Court of Justice
EU	European Union
F.2d, F.3d	Federal Reporter
F. Supp.	Federal Supplement
FTC	US Federal Trade Commission
IP	Intellectual Property
OECD	Organisation for Economic Co-operation and Development
OJ	Official Journal of the European Union (Official Journal of the European Communities prior to 1 February 2003)
SSNIP	Small but significant non-transitory increase in price
TTA Guidelines	Guidelines on the application of Article 81 of the EC Treaty to technology transfer agreements
TTBER	Transfer of Technology Block Exemption Regulation
US	United States, United States Report
USC	United States Code

Table of legislation, cases and other official documents

UNITED STATES

Supreme Court Cases by Year

1. Introduction

1. BACKGROUND

The antitrust law of tying arrangements was born amid policy disputes about the proper scope of patent coverage. Initially, the complaints were that the defendant was attempting to expand the scope of its patents monopoly by requiring the licensee to purchase unpatented complementary products as a condition of obtaining a license to the patent. The Supreme Court's conclusion that tying arrangements in the presence of tying product power are unlawful 'per se' expressed a policy judgment that severely limited the scope of the intellectual property grant, particularly because the rule was applied to IP owners who in fact lacked significant market power.[1]

The US IP law has since changed its approach to tying and now sees it as *legal per se* and importantly recognises that tying requires significant market power to be harmful and that this market power does not stem exclusively from the IP right itself. US antitrust law appears to be moving in the same direction. However, it is doubtful whether the same can be said for EC competition law. In a time when globalisation and increased cooperation between states, not least in the sphere of competition law, are becoming more significant for the world economy, harmonisation or at least a similar mindset on how to treat multi-national corporations and their actions in the market place plays a crucial role.

In the last couple of years there have been two major developments which have sparked the writing of this book. First, two major and very similar cases involving one multi-national company, Microsoft, have gone through the courts in both the US and the EC, both revolving around the conduct of tying. The cases have provided an excellent opportunity for a comparison of the approaches applied to tying arrangements on both sides of the Atlantic and highlight the flaws and failings of the current approaches to tying, in particular the wooden separate product test. The saga however, does not end here; the EC Commission has filed preliminary charges against Microsoft in yet another tying case, and in the US Microsoft continues to be monitored.[2]

---

[1]   Hovenkamp, Herbert, 'IP ties and Microsoft's rule of reason', *The Antitrust Bulletin*, Summer–Fall 2002, 369, at 372.

[2]   Keizer, Gregg, 'Update: EU hits Microsoft with new antitrust charges',

The book therefore comes at a time when the understanding of tying and its effect upon the markets is crucial, especially with the increasingly rapid development of technologically advanced and complex products. There is an urgent need to develop competition rules which encourage innovation but retain the delicate balance between the protection of innovation and ensuring a competitive market.

Recent events and challenges presented by unilateral conduct to competition enforcement have prompted both the European Commission (the Commission) and the US Department of Justice (DoJ) to review the approach taken under Article 82 of the EC Treaty and Section 2 of the Sherman Act,[3] which is the second important development influencing the writing of this book. The review of Article 82 is driven by a desire to align the legal analysis of Article 82 with that of the recent modernisation of Article 81 EC and the EC Merger Regulation[4] and develop a more economic approach, which can both accommodate recent economic thinking by being more realistic or true to the reality of the market place, and contain practical and workable rules.[5] The Section 2 Report of the US Department of Justice highlights similar concerns; namely the search for 'clear, objective, effective, and administrable' standards, which section 2 case law so far has been unable to achieve of its own accord.[6]

Computerworld, 16 January 2009 (www.computerworld.com/action/article.do?command=viewArticleBasic&articleId=9126221); Waters, Richard, 'EU launches new Microsoft antitrust probe', *Financial Times*, 14 January, 2009 (www.ft.com/cms/s/0/2ce90532-c2c1-11dc-b617-0000779fd2ac,_i_email=y.html) and US Department of Justice, Antitrust Division (www.usdoj.gov/atr/cases/ms_index.htm).

[3]   The European Commission published its 'Discussion Paper on Article 82 of the EC Treaty' in 2005: available at http://ec.europa.eu/comm/competition/antitrust/art82/discpaper2005.pdf. and its *Article 82 Guidance* in December 2008: Guidance on the Commission's Enforcement Priorities in Applying Article 82 EC Treaty to Abusive Exclusionary Conduct by Dominant Undertakings, Brussels, 3 December 2008 (http://ec.europa.eu/competition/antitrust/art82/guidance.pdf). The US Department of Justice (DoJ) in June 2006 commenced hearings on the Sherman Act Section 2, which culminated in the publication of a report on the same matter in September 2008: 'Competition and Monopoly: Single-Firm Conduct Under Section 2 of the Sherman Act (2008)', available at www.usdoj.gov/atr/public/reports/236681.htm. Note the report was controversially withdrawn by the DoJ in May 2009, see press release 09-459 'Justice Department Withdraws Report on Antitrust Monopoly Law' 11 May 2009 (www.usdoj.gov/opa/pr/2009/May/09-at459.html). The Canadian Competition Bureau followed suit by also launching an update of its 'Enforcement Guidelines on the Abuse of Dominance Provisions' (www.competitionbureau.gc.ca/eic/site/cb-bc.nsf/eng/h_00511.html).

[4]   Council Regulation (EC) No 139/2004 of 20 January 2004 on the control of concentrations between undertakings [2004] OJ L24/1.

[5]   Commission 'Article 82 Review' (http://ec.europa.eu/comm/competition/antitrust/art82/index.html).

[6]   *Section 2 Report*, *supra* n. 3, p. vii.

Both competition authorities and scholars highlight the need for a modernisation of the approach to unilateral conduct, and this includes in particular a more economic approach – one that moves away from *per se* illegality of certain types of conduct. The book will contribute to this modernisation debate in relation to tying and will offer a solution.

2. TYING

Tying is the selling or licensing of a product or service made conditional on the sale or licensing of another (related) product. It is a common and well-established business strategy applied by large and small companies, in competitive markets as well as in oligopolies, and monopolies. The reasons behind tying are plentiful: cost reduction, quality improvement, and reducing price inefficiencies; but it can also be used to foreclose markets, undercut rivals, mitigate competition, obscure prices and develop new products through technological integration.

Consumers are faced with tying of products or services on a daily basis, in the form of cars with air conditioning and stereos; computers with operating systems and internet browsers; telephone and broadband connections.

Industries and products are developing rapidly, offering increasingly more technologically advanced and complex products. Many are inter-linked, consisting of add-on features, for instance, the mobile phone, which in turn led to the development of the camera phone, music player and internet browser, all contained in one small phone.

Early US antitrust law treated tying as severely as price fixing, market division and other forms of horizontal restraints, and held it to be illegal *per se* despite tying in nature being a vertical restraint and without clear evidence of actual harm. However, in later cases the US courts required there to be market power before deeming the tying arrangement illegal, and a four-step test was developed to assess tying. The four steps were (1) the need for market power, (2) the presence of two separate products, (3) the finding of (potential) anti-competitive effects, and (4) the lack of any objective justifications for the tying arrangement.

Yet the US courts have continued to label tying as illegal *per se*, and it has remained one of the main concerns of the US antitrust authorities. In comparison, tying has only recently entered into the sphere of greater attention from the EC competition authorities,[7] although EC case law demonstrates an

---

7   Waelbroek, Denis, 'The compatibility of tying agreements with antitrust rules: a comparative study of American and European rules' [1987] *Yearbook of European Law* 39, at 39.

equally harsh treatment of tying arrangements and a similar four step-test has been deployed. In other words, once a dominant company has been found to tie products or services together, it has been found to infringe competition law without much consideration being given to the effects the tying conduct has on the market and the competitors.

The recent *Microsoft* cases[8] on both sides of the Atlantic have resulted in renewed attention being given to tying as unilateral conduct and have sparked a debate amongst both lawyers and economists questioning the legal approach taken to tying. In particular, many economists have argued that tying is in fact not the crude business strategy that it has been made out to be in the courts, but can in certain circumstances be an important tool for innovation.[9] The argument follows the idea that technological integrated (tied) products are not necessarily anti-competitive but in fact a form of innovation – a way for companies to innovate and introduce new products into the market. There is therefore a clear gap between the current case law approach and economic thinking in respect of tying which is worth exploring. Consequently, many scholars have argued for a more lenient treatment of tying in case law. This book will lend itself to the same line of arguments.

In the *Microsoft* cases, Microsoft in the US was accused of tying its Internet browser, Internet Explorer, to its operating system, Windows; similarly, in the EU Microsoft was accused of tying its media player, Windows Media Player, to Windows. Although the four-step approach to tying was applied on both sides of the Atlantic and on the facts the cases were very similar, the outcomes differed significantly. In the US, the Court of Appeals remanded the tying issue to the lower court for a *rule of reason* assessment, whereas in the EU the Court of First Instance (CFI) found Microsoft guilty of tying and required it to offer a version of Windows without its Windows Media Player.

---

[8]   Case COMP/ C-3/37.792 *Microsoft*, [2005] 4 CMLR 965, Case T-201/04 *Microsoft Corp. v Commission*, Judgment of 17 September 2007; *United States v Microsoft Corp.* (*US Microsoft II*), 147 F.3d 935 (D.C. Cir. 1998) and *United States v Microsoft Corp.* (*US Microsoft III*), 253 F. 3d 34 (D.C. Cir. 2001).

[9]   Ahlborn, Christian, Evans, David S., and Padilla, A. Jorge, 'The antitrust economics of tying: a farewell to *per se* illegality', *Antitrust Bulletin*, Spring–Summer 2004, 287; Evans, David S., and Salinger, Michael, 'Why do firms bundle and tie? Evidence from competitive markets and implications for tying law' (2005) 22 *Yale Journal on Regulation* 37; Hylton, Keith N., and Salinger, Michael, 'Tying law and policy: a decision-theoretic approach' (2001) 69 *Antitrust Law Journal* 469; and O'Donoghue, Robert and Padilla, A. Jorge, *The Law and Economics of Article 82 EC* (Hart Publishing, Oxford and Portland, Oregon, 2006). Note the more cautious approach by Kühn, Kai-Uwe, Stillman, Robert and Caffarra, Christina, 'Economic theories of bundling and their policy implications in abuse cases: an assessment in light of the Microsoft Case' (2005) *Vol. 1, no. 1, European Competition Journal* 85.

Importantly, the *Microsoft* cases demonstrated a difference in the understanding of products and what constitutes an objective justification. In particular, the cases raised the question of when a product ceases to be a product and instead becomes a component of another product. Yet neither the US nor the EC case provided an answer, despite being of great significance to the understanding of tying. The book will explore this question in greater detail and seek to provide an answer.

The *US Microsoft* case also marked a (potential) turning point in US case law towards a more lenient approach to tying, and in particular technological integration. Interestingly, this seems to have been mimicked in the *Section 2 Report* from the US Department of Justice.[10] In comparison, the *EC Microsoft* case left little sign of a similar turnround for future EC case law.

The *Microsoft* cases exhibited the concerns of the competition authorities to ensure effective competition in high-technology industries and fast moving markets, but also showed that the current approach to tying is ill-equipped to handle more advanced forms such as technological integration.

3. OUTLINE OF THE BOOK

The book will firstly assess the current approach to tying arrangements as a form of unilateral behaviour in both the EU and the US. Secondly, it will discuss the scope for improvements by assessing the four-step tests for tying which have been developed under both EC and US competition law. The book will outline a proposal for an adapted test for tying based on the current EC competition law approach, but will also address its application to US antitrust law. The book will apply three different perspectives: an economic perspective, a competition law perspective, and an IP law perspective. The development of competition law demonstrates that it is highly reliant upon economic theories and principles, and thus an assessment of tying would be incomplete without a discussion of the underlying economic thinking behind it as a business strategy. The economic perspective will also provide a reference point in the book for why there should be a change in the legal approach to tying.

The IP law perspective is included for two reasons: firstly, the US approach to tying has its origins in the Patent Misuse Doctrine and, secondly, in particular in the EU, most of the tying cases, where the tying was found to be abusive, involved some form of IP right.

Although many scholars have acknowledged the link between US antitrust and IP law in relation to tying, none seems to have taken this further. With the

---

[10] *Section 2 Report, supra* n. 3, 88–89.

inclusion of the IP law perspective this book seeks to do so by asking the following questions: firstly, whether tying of IP rights or IP right protected products is treated more harshly than other forms of tying arrangements; secondly, whether we can learn something from the IP law approach to tying – does the IP law provide an alternative model to tying?

The starting point, in Chapter 2 is an overview of economic literature and economic arguments applicable to tying arrangements in an attempt to draw lessons from the development of economic thinking on tying. The chapter will look at efficiency and strategic reasons behind tying to identify when, from an economic perspective, tying can be seen as pro-competitive and when the tying arrangement is harmful.

Chapters 3 and 4 will assess the EC competition law and US antitrust law approaches to tying respectively focusing upon Article 82 and the Sherman Act Sections 1 and 2 and the Clayton Act Section 3. The chapters will identify the problems and highlight the benefits of the current approaches by examining current case law on tying.

Chapter 5 will analyse the policy on tying arrangements adopted by US IP law. As mentioned, the US has developed a misuse doctrine mainly based on tying in IP case law. It therefore provides a useful insight into an alternative method of dealing with tying arrangements.

Chapter 6 will draw upon the lessons learned from the previous chapters, that is the economic theories surrounding tying and how they square up with the current approaches taken under EC competition law and US antitrust law, and whether any lessons can be learned or elements applied from the approach taken towards tying under the US patent misuse doctrine, to build up a suggestion for a new approach to tying under competition law. This is done by asking crucial questions, such as should tying continue to be perceived as illegal *per se*? What are the alternative solutions? And what are the consequences? Should we adopt a new test for tying? And if so what should this test look like? Which essential elements should it contain? In particular, it suggests a distinction between the treatment of contractual tying and that of technological integration, providing an all-important safe haven for innovation and product development.

Finally some concluding remarks are set out in Chapter 7, which also takes a brief look at future cases and the prospect for application of the proposed test.

2. Tying from an Economic Perspective

1. INTRODUCTION

> Every person who sells anything imposes a tying arrangement. This is true because every product or service could be broken down into smaller components capable of being sold separately, and every seller either refuses at some point to break the product down any further or, what comes to the same thing, charges a proportionally higher price for the smaller unit.[1]

Tying is a common form of business practice, yet it is also thought to produce a harmful effect on competition, especially when the tying company holds significant market power.[2] The competition laws in both the US and EC have not always appreciated that tying can be both harmful *and* offer efficiencies which enhance consumer welfare.[3]

This chapter will assess the economic incentives behind tying, some pro-, and others anti-competitive. It is hoped that gaining an understanding of the economic reasoning behind tying will provide a perspective for a more realistic legal approach to tying.

From a historical perspective, the law and economic theory of tying can be divided into three major eras, the classical, the Chicago School, and the post-Chicago School.[4] In the classical period the US courts developed a strict illegal *per se* approach to tying based on the theory that tying was an attempt to leverage market power from one market (the tying product market) into another (the tied product market).[5] From the 1950s to the 1970s the Chicago School's argument that tying should be legal was at its height, and this greatly influenced the US courts' decisions showing an increasingly narrow approach

---

[1]   Bork, Robert H. *The Antitrust Paradox: A Policy at War with Itself* (Basic Books, 1978, reprinted by The Free Press, 1993) at 378–379.

[2]   Ahlborn, Christian, Evans, David S., and Padilla, A. Jorge 'The Antitrust Economics of Tying: A Farewell to *Per Se* Illegality', *Antitrust Bulletin*, Spring–Summer 2004, 287, at 288.

[3]   Ibid.

[4]   Hylton, K.N. and Salinger, M., 'Tying law and policy: a decision-theoretic approach' (2001) 69(2) *Antitrust Law Journal* 469, p. 469.

[5]   Ibid. See also *International Salt Co. v United States*, 332 U.S. 392 (1947).

to the hitherto *per se* prohibition.[6] The outcome is the four-condition test applied today. The Chicago School theory therefore remains influential on today's legal approach to tying. The Chicago School argument is based on microeconomics and in its basic terms claims that it is not possible for one company to gain two monopoly profits.[7] Finally, the post-Chicago School theory argues that tying in fact can be anti-competitive in certain circumstances. The US courts have been less influenced by post-Chicago School arguments than the Chicago School's line of reasoning.[8] They seem to have been more influenced by economists' views on occasion than the EC Courts.

This chapter will explore the economic arguments for and against tying, in order to identify when tying is efficient and should be allowed, and when it is harmful and should be prohibited from an economic perspective. When appropriate, reference will be made to case law for comparison.

2.   DEFINING TYING

We can distinguish between different categories of tying. For instance, in his report for the DTI, Nalebuff identified three main categories of tying: bundling, tying and metering. These are further divided into subcategories such as pure bundling, mixed bundling, static and dynamic tying, pure metering and metering via tying.[9] Other economists have identified two other categories, those of commodity bundling[10] and technical bundling.[11]

Bundling has been defined as 'the practice of selling two or more differentiated products in a single package'[12] where the products are not available on

---

    6   Ibid. and *Times-Picayune Pub. Co. v United States*, 345 U.S. 594, 73 S. Ct. 872, 97 L.Ed. 1277 (1953); but compare with *Northern Pacific Railway Co. et al. v United States*, 356 U.S. 1 (1958).

    7   Nalebuff, Barry, 'Bundling, Tying, and Portfolio Effects, Part 1 – Conceptual Issues', DTI Economics Paper No 1 (2003) (www.bcrr.gov.uk/files/file14774.pdf) p. 20.

    8   Hylton and Salinger, *supra* n. 4, p. 470, and *Schor v Abbott Laboratories*, No. 05-3344 (7th Cir. 2006); however, see also Feldman, Robin C., 'Defensive leveraging in antitrust' (1999) 87 *Georgetown Law Journal* 2079.

    9   Nalebuff, *supra* n. 7, 13–17.

    10   Adams, William James and Yellen, Janet L., 'Commodity bundling and the burden of monopoly' (1976) 90 *The Quarterly Journal of Economics*, 475, p. 475.

    11   Kühn, Kai-Uwe, Stillman, Robert and Caffarra, Christina, 'Economic theories of bundling and their policy implications in abuse cases: an assessment in light of the Microsoft case' (2005) 1 *European Competition Journal* 85, p. 88 and O'Donoghue, Robert and Padilla, A. Jorge, *The Law and Economics of Article 82 EC* (Hart Publishing, Oxford and Portland, Oregon 2006), ch 9.

    12   Carboja, Jose, De Meza, David, and Seidman, Daniel J. 'A strategic motiva-

an individual basis, whereas tying is often just defined as making the sale (or price) of one product conditional upon the purchase of another product.[13] The main distinction between the two definitions is that bundling only allows for the purchase of the package of products, and hence the individual products cannot be purchased on their own, whereas in the case of tying the tying product is available on its own.[14] However, this distinction overlaps somewhat with the subcategories of pure bundling and mixed bundling. Pure bundling is where two products are sold only in the bundle, and they cannot be purchased separately. An example of this would be shoes, as these are always sold in pairs. Moreover, the products are often only available in fixed proportions, one left shoe with one right shoe.[15] In comparison, mixed bundling occurs where 'goods A and B are sold as an A-B package in addition to being sold individually. The package is sold at a discount to the individual prices.'[16] A good example is that of set menus in restaurants: it is often cheaper to opt for the set menu than to purchase the individual dishes *à la carte*.

In contrast to mixed bundling, tying only allows for either the A-B package or for A alone. If customers would like product B, they will have to purchase A as well.[17] An example of this is Windows and Internet Explorer: in order to get Windows operating system you have to purchase Internet Explorer as well, although you can download Internet Explorer without Windows.

Some scholars talk about commodity bundling as well, defined as the 'practice of package selling . . . Commodity bundles sometimes include goods that cannot be sold separately in the market place . . . Commodity bundling also occurs when firms sell the same physical commodity in different container sizes'.[18] In other words, commodity bundling is simply a broad term to cover all kinds of bundling.

Metering should be distinguished from tying applied as a metering device. Metering is defined as '. . . a form of pricing where the customer pays a per-

---

tion for commodity bundling' (1990) 38 *Journal of Industrial Economics* 283, p. 283. This definition is however, the one Nalebuff applies to 'pure bundling': Nalebuff, *supra* n. 7, 13.

[13]   Whinston, Michael D., 'Tying, foreclosure, and exclusion', (1990) 80 *The American Economic Review*, 837, p. 837.

[14]   Charles River Associates, 'Innovation and Competition Policy, Part I – Conceptual Issues', Economic Discussion Paper 3, March 2002, Report prepared for the Office of Fair Trading (www.oft.gov.uk/shared_oft/reports/comp_policy/oft377part1.pdf) (please note that Charles River Associates are now CRA International) (hereafter the Charles River Report), 78.

[15]   O'Donoghue and Padilla, *supra* n. 11, 477.

[16]   Nalebuff, *supra* n. 7, 14.

[17]   Ibid., 15.

[18]   Adams and Yellen, *supra* n. 10, 475.

use fee'.[19] Tying applied as a metering device is defined as 'requir[ing] that the customer purchase [the tied product] at an inflated price'.[20]

Technical bundling or technological integration is defined thus: '[a] special form of pure bundling arises when the two products are linked technically such that it is physically impossible for the consumer to separate them . . . it involves some form of lasting product design decision'.[21]

There is not always a clear distinction between the different categories of bundling and tying or rather the definitions somewhat overlap.[22] For instance, Carboja *et al.* state that 'commodity bundling . . . is a special case of tie-in sale'[23] in contrast to Nalebuff, who defines static tying 'as a special case of mixed bundling' and dynamic tying 'as a dynamic form of pure bundling'.[24]

That said, from the tying categories discussed above, it is possible to identify four different tying scenarios – the term in brackets will be the term applied for the specific definition throughout the book:

(1) The two products can only be purchased together (pure bundling).
(2) The two products can be purchased together and one of the products can be purchased on its own (tying).
(3) The two products can be purchased together and both products can be purchased on their own (mixed bundling).
(4) The two products are technically integrated (technological integration).

It is clear that the types of tying applied, whether pure bundling, tying or technological integration, will create different effects upon the market and therefore cause different levels of harm and efficiencies.[25] It indicates that the different subcategories of tying identified above can play an important role in whether the tying conduct is or should be unlawful. Secondly, the subcategories of the tying definition can also indicate the purpose of the tie. The choice of tie or bundle will to some degree determine the anti-competitive effects, as for instance mixed bundling, which offers the products in a package

---

19   Nalebuff, *supra* n. 7, 16.
20   Ibid., 17.
21   Kühn, Stillman, and Caffarra, *supra* n. 11, 88.
22   See Bishop, Simon and Walker, Mike: *The Economics of EC Competition Law, Concepts, Application and Measurements* (Sweet and Maxwell, London 2002), para. 6.54.
23   Carboja, De Meza and Seidman, *supra* n. 12, 283.
24   Nalebuff, *supra* n. 7, 15 compare with Ahlborn, Evans and Padilla, *supra* n. 2, 287–341.
25   Economists have proved that under certain conditions mixed bundling will create higher profits than pure bundling: see Adams and Yellen, *supra* n. 10.

or individually, will cause less harm than tying or pure bundling where the products are offered only as a package.[26]

3.   THE PURPOSES OF TYING

Two major theories have persistently held tying to be as an anti-competitive strategy; one is the theory of leveraging; the other is price discrimination theory.[27] The two will be dealt with separately below. However, the former is the more relevant for this book, as case law shows that this has been seen as the major motivator of tying and also the one causing most concern.

The reasons behind tying, identified by the scholars associated with the Chicago School and later verified and expanded by others, are numerous.[28] For instance, tying can be applied as a means of quality control, distribution of franchisors' products, protection of trade secrets, licence royalty substitute, and metering. Tying can also be employed to aid entry into a market or achievement of economies of scale, and finally tying can technically integrate new products.[29] As highlighted above it can also be a means to expand into an adjacent market, keeping potential competitors out of a market and preserving a monopolist's market power. That said, Ahlborn *et al.* state '[tying] is so common in competitive markets that it must provide efficiencies'.[30] In contrast, Kühn *et al.* argue that the fact that tying or bundling can induce efficiencies has been 'abused for sweeping claims about policy'.[31] They continue by highlighting three erroneous arguments:

> First, efficiencies that are attributed to bundling often can be achieved without bundling. Secondly, efficiency claims for technical bundling are given with reference to examples where this occurs. At the same time the abundance of examples in which such potential joint production does not occur is suppressed. Thirdly, theories are misinterpreted to lead to efficiency effects of bundling when these theories do not imply such conclusion at all.[32]

---

26   The Charles River Report, *supra* n. 14, 78.
27   Evans, David S, and Salinger, Michael, 'Why do firms bundle and tie? Evidence from competitive markets and implications for tying law' (2005) 22 *Yale Journal on Regulation* 37, p. 48.
28   Ibid., Nalebuff, *supra* n. 7, and the Charles River Report, *supra* n. 14.
29   Hovenkamp, Herbert, Janis, Mark D. and Lemley, Mark A., *IP and antitrust, an analysis of antitrust principles applied to intellectual property law* (Aspen Law & Business, New York, 2007), vol. I, Section 21.5b2.
30   Ahlborn, Evans and Padilla (2004), *supra* n. 2, 289.
31   Kühn, Stillman and Caffarra, *supra* n. 11, 106–107.
32   Ibid., 107.

Table 2.1   Efficiency and strategic reasons for tying

1 Efficiency reasons	2 Strategic reasons
1 Cost reduction	1 Leveraging 　　a)　Market foreclosure 　　b)　Protection of the tying 　　　　product market 　　c)　Tying applied to undercut 　　　　rivals' Price
2 Quality improvement 　　a)　Quality assurance 　　b)　Reduction in search costs 　　c)　Product improvement	2 Mitigation of competition
3 Reducing pricing inefficiencies 　　a)　Price discrimination 　　b)　Double marginalization	3 Gaining competitive advantage 　　a)　Creating network 　　　　externalities 　　b)　Technological integration
	4 Price obfuscation
	5 Technological integration
	6 Increasing R&D

Nalebuff has identified several reasons for a company to apply tying, which he has divided into two main categories, namely efficiency and strategic reasons.[33]

These categories differ from those adopted by others,[34] by virtue of the fact that they create a clear distinction between efficiency benefits of tying and anti-competitive effects of the conduct, whereas in fact these categories over lap.[35] Hence, a monopolist may reduce pricing inefficiencies by tying, but at the same time enhance its market share due to the fact that consumers cannot differentiate between the total price of the tied package and the price or cost of the individual products.[36] Most scholars argue that these economic and

---

[33]   The table, although based on Nalebuff's findings, (Nalebuff, *supra* n. 7, 18), has been expanded by the author to include other scholars' findings: see O'Donoghue and Padilla, *supra* n. 11, pp. 480–491.

[34]   Kühn, Stillman and Caffarra, *supra* n. 11.

[35]   Nalebuff, *supra* n. 7, 18.

[36]   Ibid. and Hylton, Keith N., *Antitrust Law, Economic Theory & Common Law Evolution* (Cambridge University Press, Cambridge, 2003), 282.

competitive benefits should be taken into consideration by the competition authorities instead of the illegal *per se* policy or before refusing a dominant company the chance to tie products, because the efficiencies gained by the tie can often outweigh any competitive harm.[37]

The analysis below follows the outline of the table above. The first section discusses the different efficiencies and welfare-increasing effects that economists have identified in tying but which have not always been recognised by the competition authorities. The second section deals with the strategic reasons behind tying, those often seen as anti-competitive by the competition authorities.

4. EFFICIENCY REASONS

4.1. Cost Reduction

Tying can result in reduction in production, distribution, marketing, licensing costs, and create economies of scale for the company, as a manufacturer may benefit from producing and selling two products together rather than selling them individually. For instance, the manufacturer can simply save time and money by packaging the two products together. The most basic example is that of shoes or socks: it is beneficial for both manufacturer and customers to sell and purchase them together in pairs as indeed most people want matching pairs. The fact that these products are sold bundled has not raised particular concerns or questions. Salinger explains that it is more profitable to bundle because the bundle will lower the costs 'when demands for the components are highly positively correlated and components costs are high'.[38] Taking shoes as an example: if most consumers only purchased the pair (or none), then even a small saving in costs would be an incentive for a monopolist to sell the shoes together. However, Kühn *et al.* argue that this is a joint production fallacy – just because the manufacturer benefits from producing the products together does not require him to sell them together.[39]

> General claims on efficiencies generated by bundling ignore the point that any efficiency enhancing bundling must be driven by preference for joint consumption (i.e.

---

[37] See Hovenkamp, Janis and Lemley, *supra* n. 29, 21–28; Bork, *supra* n. 1, 381; Hylton and Salinger, *supra* n. 4, pp. 470–471 and Lind, R.C. and Muysert, P., 'Innovation and competition policy, challenges for the new millennium' (2003) 24 *European Competition Law Review* 87, p. 90.

[38] Salinger, Michael A., 'Graphical analysis of bundling' (1995) 68 *The Journal of Business* 85, p. 98.

[39] Kühn, Stillman and Caffarra, *supra* n. 11, 107.

complementarity between the goods). But even when there is a preference for joint consumption, it is not necessary for complements to be sold on a bundled basis, unless this action itself leads to real economies.[40]

Kühn *et al.*'s argument is dismissed by O'Donoghue and Padilla. They highlight that although Kühn *et al.* are correct in their initial assumption that tying does not create savings in production costs, they have overlooked the risk of diseconomies of scale by offering several separate products instead of the package.[41] Evans and Salinger clarify that tying is often applied in competitive markets as an efficiency device to cut fixed costs.[42] Technically, the companies are simply refusing to sell the tied product without the tying product. Although in principle this limits consumers' choice, it must be recognised that it is not always possible for companies to offer certain products at profitable prices that consumers would be willing to pay.[43]

Equally from a marketing perspective it can be advantageous to tie products or services together. For instance, cable television markets often consist of bundled products and services, in particular in recent years with the rapid development and expansion of the Internet. Hence companies like Sky and Virgin Media offer Internet access, pay TV and phone connections tied together.

Taking the cable television market as an example, three crucial characteristics demonstrate that bundling is demand-based in that market: (1) historically, the market was monopolised; (2) there is diversity in the mix of offerings for the basic packages which results in the uncorrelated and negatively correlated values of the different channel packages; and (3) there are low marginal costs.[44]

Another form of cost saving is reduction in transaction costs, which will directly benefit the consumer in the form of a lower package price.[45] However, the consumer may also make transaction costs savings as, by choosing the bundle, the consumer will not have to search for the best available combinations of the different component goods.[46] As an example, computers are now frequently sold in one package consisting of hardware, software, monitor, keyboard and mouse. In the early life of the personal computer, each component often had to be purchased individually.

---

40   Ibid.
41   O'Donoghue and Padilla, *supra* n. 11, 481.
42   Evans and Salinger, *supra* n. 27, 85.
43   Ibid.
44   Salinger, *supra* n. 38, footnote 17.
45   Charles River Report, *supra* n. 14, 81.
46   Ahlborn, Evans and Padilla (2004), *supra* n. 2, 320.

4.2. Quality Improvement

4.2.1. Quality assurance

Closely related to cost savings is the issue of quality assurance. Not only may tying certain products together create economies of scale for the manufacturer, it may also improve the quality of the products.[47] If a bicycle were to be purchased in separate parts which the consumer herself had to assemble, there would be a risk that the quality would not be as good as if the manufacturer had sold the bicycle as a bundle and had assembled it prior to purchase. Naturally, this applies to a greater extent the more technologically sophisticated the products are.[48] Hence tying products together can benefit both manufacturer and consumer.

Tying is an efficient means of controlling quality and protecting trade marks or trade secrets in franchising and leasing agreements. A basic example is photocopiers and cartridges, where the manufacturer requires the use of its own cartridges as a condition of the warranty on the photocopier.[49] In the event that something goes wrong the customer will naturally blame the manufacturer of the photocopier, although the fault may lie with the cartridge manufacturer. Hence tying the cartridge to the photocopier benefits both customer and manufacturer. The former has only one manufacturer to deal with and thus the latter will be compelled to honour its warranty, but can also easily monitor its own product quality and protect its trade mark reputation.[50]

However, this justification can be abused relatively easily. It is a simple matter to argue that such effects occur but difficult to assess the genuineness of such an argument.[51] Quality control claims have been accepted by the courts both in the EU and the US, but have required a high standard of proof.[52]

The US Supreme Court stated in the legendary *Jefferson Parish* case that 'contractual quality specifications are generally sufficient to protect quality without the use of tying arrangement',[53] thus indicating that tying should be seen as the last resort for quality control. This is much in line with the EC

---

[47]   Ibid., 321–322.

[48]   Ibid.

[49]   The example is borrowed from Hovenkamp, Janis and Lemley, *supra* n. 29, 21–8 to 21–9.

[50]   Charles River Report, *supra* n. 14, 81.

[51]   Ibid., 82.

[52]   *IBM Corp. v United States*, 298 U.S. 131 (1936) at 139–140; *International Salt supra* n. 5, and Case T–30/89 *Hilti v Commission* [1991] ECR II–1439, [1992] 4 CMLR 16, paras 102–114.

[53]   *Jefferson Parish Hospital District No. 2 v Hyde*, 466 U.S. 2, 6 n. 42 (1984).

Courts' attitude.[54] This was also the view of the Supreme Court in the much earlier case of *Standard Stations* in which it found that the only situation where 'protection of good will may necessitate the use of tying clauses is where specifications for a substitute would be so detailed that they could not practicably be supplied'.[55] These two statements do not take the importance of numerous agency and free rider problems into consideration and consequently restrict some of the benefits tying can offer to limit the problems the franchisor faces.[56] For instance, the franchisor can experience free rider problems because franchisees do not bear the full cost of relying on lower quality products, which therefore results in a superfluous use of inferior products and ultimately damages the franchisor's name/chain.[57] This is also linked to applying tying as a means of ensuring the quality guarantee that attaches to a trade mark or a manufacturer's brand.[58]

4.2.2. Reduction in search costs

Tying can reduce the costs of searching for the most appropriate combination of products working together. This is particularly essential for more complex products. 'The widespread use of bundled software is itself a function of better technology . . .'.[59] In the early days of the computer era, software technology such as toolbars, power management and modem support were all offered as stand-alone products. Today they are all offered as part of the operating system.[60] This development was also seen with the stereo – originally consumers purchased record player (later CD player), tape recorder, radio and amplifier separately. Nowadays all are combined in a small, unified box. The tying of these complex components is not just the result of innovation and better technology, but also a response to consumer demand.[61]

---

[54]   See Case T–30/89 *Hilti*; C–333/94 *Tetra Pak International SA v Commission* (*Tetra Pak II*) [1996] ECR I–5951, [1997] 4 CMLR 662; and 'Guidance on the Commission's Enforcement Priorities in Applying Article 82 EC Treaty to Abusive Exclusionary Conduct by Dominant Undertakings', Brussels, 3 December 2008 (http://ec.europa.eu/competition/antitrust/art82/guidance.pdf) (the *Article 82 Guidance*), 18.

[55]   *Standard Oil Co. of California v US*, 337 U.S. 293, 69 S.Ct. 1051, 93 L.Ed. 1371 (1949) at 306.

[56]   Hovenkamp, Janis and Lemley, *supra* n. 29, pp. 21–29.

[57]   Ibid.

[58]   Ibid.

[59]   Evans, D.S., Padilla, A.J., and Polo, M., 'Tying in platform software: reasons for a *rule-of-reason* standard in European competition law' (2002) 25 *World Competition* 509, p. 509.

[60]   Ibid.

[61]   Ibid.

In comparison, note the example of the IKEA flat-pack which Kühn *et al.* give; in this example both consumer and manufacturer benefit from the fact that the consumer herself will have to assemble the product – the manufacturer obtains certain efficiencies and the consumer a lower price.[62] Kühn *et al.* warn that '[basing] strong conclusions on casual empiricism is a dangerous exercise. There is no basis for going from the fact that product integration sometimes produces consumer benefit to the conclusion that product integration almost invariable produces consumer benefits.'[63] With this statement, Kühn *et al.* were heavily criticising Alhborn *et al.*, although the latter do point out that the benefit of product integration increases with the level of technological sophistication of the product.[64] Hence, the assessment of whether efficiencies are gained by bundling or product integration should be based on the level of technological sophistication. Thus, the more technologically advanced a product is, the more beneficial it would probably be for the consumer to receive it as an all-in-one pack. That said, returning to the computer example, it was only later when the computer had become a common household phenomenon that computer packs were made readily available. This suggests that product integration or bundling is more likely to occur after a new product has been available in the market for a certain period of time. This is certainly true for technologically advanced products. The integration is most likely to occur when the products become more common, indicating a consumer need for integration to make the product user-friendly. The question to ask is whether product integration occurs as a result of consumer preference and natural progression or as a result of the manufacturer's ability to foresee that, by tying products together or creating product integration, it will gain a competitive advantage in the future. Evans *et al.* argue that it is the former.[65] It is, however, most likely a combination of both, as not only will consumers benefit from the integration, but the market also expands by integration or rather by making the products more user-friendly, thereby benefiting the manufacturer.

4.2.3. Product improvement

A bundle of products can be more valuable than the sum of the individual products because the combination creates more benefits for consumers than the individual products added together would be able to do.[66] An example is mobile phones, which allow the consumers to phone, text, take pictures and

---

[62] Kühn, Stillman and Caffarra, *supra* n. 11, 107.
[63] Ibid.
[64] Ahlborn, Evans and Padilla (2004), *supra* n. 2, 321–322.
[65] Evans, Padilla and Polo, *supra* n. 59, 509.
[66] O'Donoghue and Padilla, *supra* n. 11, 482.

videos and surf the Internet. The ability to send a picture without having to go through another machine, as would be the case with an ordinary camera, provides benefits that could not be achieved from using a phone and a camera. The product improvement argument was assessed to a great extent in *US Microsoft II*.[67] Microsoft argued that there were consumer benefits from the integration of its Windows operating system with its Internet Explorer browser, in the sense that they have the same format and new windows need not be opened to use some applications. In a minority opinion, Judge Wald noted some elements of the Windows/Internet Explorer integration were 'clearly greater than the sum of its parts'.[68]

4.3. Reducing Pricing Inefficiencies

A reduction in pricing inefficiencies can be achieved either by applying tying as a price discrimination device or employing it as a means of eliminating double marginalisation problems.

4.3.1. Price discrimination

Price discrimination was offered by followers of the Chicago School arguments as an alternative explanation to leveraging as the main motivator for tying.[69] The theory that tying can be deployed as a price discrimination device can be divided into two explanations. One, first pursued by Stigler,[70] revealed that tying enables companies to profit from differences in value that the consumers place on the separate products.[71] This is particularly the case where consumers are heterogeneous[72] and there are low or no marginal costs, as this will increase demand without increasing the costs.[73] Bowman developed a

---

[67]   *US v Microsoft Corp.* (*US Microsoft II*) 147 F.3d 935 (1998) at 950–952.

[68]   Ibid., at 961, see also Page, W. and Lopatka, J., 'The dubious search for "integration" in the *Microsoft* trial' (1998–99) 31 *Connecticut Law Review* 1251, p. 1268 and footnote 97.

[69]   Choi, Jay Pil, 'Tying and innovation: a dynamic analysis of tying arrangements', revised May 2002 (www.msu.edu/~choijay/Tying.pdf, p. 5 and Ahlborn, C., Denicolò, V., Geradin, D., and Padilla, A.J., 'DG Comp's Discussion Paper on Article 82: Implications of the Proposed Framework and Antitrust Rules for Dynamically Competitive Industries', 31 March 2006 (http://papers.ssrn.com/sol3/papers.cfm?abstract_id=894466), p. 40.

[70]   Stigler, George, *The Organization of Industry* (Richard D Irwin, Inc. Homewood, Ill., 1968), 165–170.

[71]   Evans and Salinger, *supra* n. 27, 49.

[72]   Ibid., 50.

[73]   Bakos, Y and Brynjolfsson, E, 'Bundling information goods: pricing, profits and efficiency' (1999) 45 *Management Science* 1613, p. 1617.

second explanation.[74] He identified that by tying a consumable product to a durable product, companies could charge more to consumers who had a greater demand for the durable product.[75]

Tying can also be used as a form of metering device to create price discrimination or as a form of licence royalty. In this kind of context, the tie-ins are pro-competitive even though they allow the tying company to gain higher returns than it would without price discrimination.[76] For instance, IBM applied tying as a metering device.[77] By tying its punch cards to its punch machines IBM was able to extract higher prices from those customers who were using the punch machines more intensively then those who were not.[78]

Price discrimination by tying can be an effective way of recovering fixed costs, in particular in markets where these are high.[79] The Charles River report notes that, unlike tying with cost reduction as its motivation, employing tying as a device for price discrimination requires market power in the tied product market.[80] However, when price discrimination is applied in conjunction with tying or metering and in markets where there are high fixed costs, network effects, or economies of scale, tying may have a positive effect upon welfare overall.[81] For example in computer games markets with gaming brands such as Nintendo, Sega and Sony which depend on committed gaming consoles and separately sold games, efficiency may be increased by cross-subsiding costs from the tied product to the tying product.[82] Software products have high fixed costs and low marginal costs, and therefore significant economies of scale.[83] By cross-subsiding the low marginal costs from software to hardware, where the marginal costs are much higher, the marginal costs can be reduced and thereby entice customers which otherwise would have been reluctant to buy into the network.[84] The additional customers will increase the installed base hardware and the attractiveness of the network, leading to lower prices and more software being produced for the hardware.[85]

---

[74]   Bowman, Ward S., 'Tying arrangements and the leverage problem' (1957) 67 *The Yale Law Journal* 19.

[75]   Evans and Salinger, *supra* n. 27, 49.

[76]   Hovenkamp, Janis and Lemley, *supra* n. 29, 21–12.

[77]   *IBM Corp.* at 134.

[78]   Evans and Salinger, *supra* n. 27, 50.

[79]   Charles River Report, *supra* n. 14, pp. 82–83.

[80]   Ibid.

[81]   Ibid.

[82]   Ibid.

[83]   Ibid.

[84]   Ibid., 83.

[85]   Ibid.

This kind of practice is also found in other industries such as those for mobile phones and network operators.[86]

In these situations, it is important to distinguish between fixed proportion tying arrangements, where there is a one to one ratio between the tied and tying product, and variable tie-ins, where the amount of the tied product can fluctuate depending on the customer's need.[87] An example of a fixed proportion tie could be a car and car stereo; only one stereo is needed per car. However, a tie between a photocopier and paper is a variable proportion tying arrangement, since it depends on how much the customers photocopy. However, in a monopoly a tie of perfect complements will not bring any efficiencies.[88] The variable proportion is effective as a price discrimination device. Under US law, the Robinson–Patman Act[89] outlaws certain forms of price discrimination, but tying arrangements applied as price discrimination devices in principle do not violate it.[90] This is because the Act requires the same product to be sold at two different prices, whereas in a tying arrangement discrimination arises because the tying and tied products are applied in different quantities to different customers.[91]

The equivalent rules in Europe to the Robinson–Patman Act are Articles 81(d) and 82(c), which prohibit the application of 'dissimilar conditions to equivalent transactions with other trading parties, thereby placing them at a competitive disadvantage'.[92] Economists argue that price discrimination, like tying, can be both harmful and beneficial to competition. However, they conclude that price discrimination is mainly beneficial.[93] This means that tying arrangements which facilitate price discrimination can in principle be pro-competitive and welfare enhancing.[94] This is essentially the approach

---

[86]   Ibid.

[87]   Hovenkamp, Janis and Lemley, *supra* n. 29, 21–12.

[88]   Kühn, Stillman and Caffarra, *supra* n. 11, 91.

[89]   The Robinson–Patman Act of 1936 (Anti-Price Discrimination Act) 15 U.S.C. 13 orignally an amendment to the Clayton Act, section 2.

[90]   Hovenkamp, Janis and Lemley, *supra* n. 29, 21–12, footnote 15.

[91]   Ibid.

[92]   See also *Article 82 Guidance*, *supra* n. 54, 18.

[93]   Van den Bergh, Roger J. and Camesasca, Peter D., *European Competition Law and Economics: A Comparative Perspective* (2nd edn., Thomson Sweet & Maxwell, London, 2006), 270–271, Motta, Massimo, *Competition Policy – Theory and Practice* (Cambridge University Press, New York, 2004), p. 463, who notes that the welfare implications of tying are ambiguous; Tirole, J., *The Theory of Industrial Organisation* (The MIT Press, Cambridge, Mass., 1978 (reprinted 2003)), 146–148; Ahlborn, Denicolò, Geradin, and Padilla, *supra* n. 69, 40; Ahlborn, Evans and Padilla (2004), *supra* n. 2, footnote 119; Bork, *supra* n. 1, 381; and Hovenkamp, Janis and Lemley, *supra* n. 29, 21–13.

[94]   Ahlborn, Evans and Padilla (2004), *supra* n. 2, footnote 119.

adopted in the US. In contrast, in Europe price discrimination remain mostly illegal *per se*.[95]

Kühn *et al.* argue that in welfare terms there is a trade-off between two consumer groups when bundling is applied as a means of price discrimination.[96] Some buyers are forced to purchase a second product they do not value as highly as the first, whereas the other consumer group will benefit from the bundle as they value the products equally highly.[97] Nalebuff is not convinced that price discrimination is the most significant motivator for tying, because products which are tied together are often positively correlated in value, and this therefore suggests that there are incentives behind bundling other than price discrimination.[98] Equally, Nalebuff states that it is a false assumption to believe that price discrimination is efficient; only perfect price discrimination is efficient, and in practice perfect price discrimination does not exist.[99] This is also the opinion of Kühn *et al.*, who note that the welfare effects generated from price discrimination ties are unpredictable.[100] This is due to at least two facts: first, companies tend to have limited possibilities for price discrimination and, second, no monopoly is perfect. Hence, there will always be some form of substitute available.[101] For instance, the Royal Mail had a monopoly for many years in the United Kingdom, as did many other national postal delivery companies worldwide. However, other means of communicating, such as phone, email and fax, were available. Customers will attempt to avoid price discrimination. An example is the airlines charging less for tickets with no frills and purchased months in advance of the travel date in an attempt to discriminate between leisure and business travellers. The natural outcome is of course that the business travellers will buy the no-frills tickets to obtain the cheaper price. The result, Nalebuff argues, is 'large social losses, well above the incremental gains from the extra demand that arises at low prices'.[102]

4.3.2. The double marginalisation problem

In theory, competition law aims to guide the market towards a perfect competition equilibrium, where there are in principle no social losses. The demand curve is horizontal (see Figure 2.1) and the firms are price-takers, because of large numbers of sellers and buyers and homogenous products. The firms in these markets will adjust costs to the price given.

---

[95] Ibid.
[96] Kühn, Stillman and Caffarra, *supra* n. 11, 90–91.
[97] Ibid.
[98] Nalebuff, *supra* n. 7, 37.
[99] Ibid., p. 77.
[100] Kühn, Stillman and Caffarra, *supra* n. 11, 91.
[101] Nalebuff, *supra* n. 7, 77.
[102] Ibid., 78.

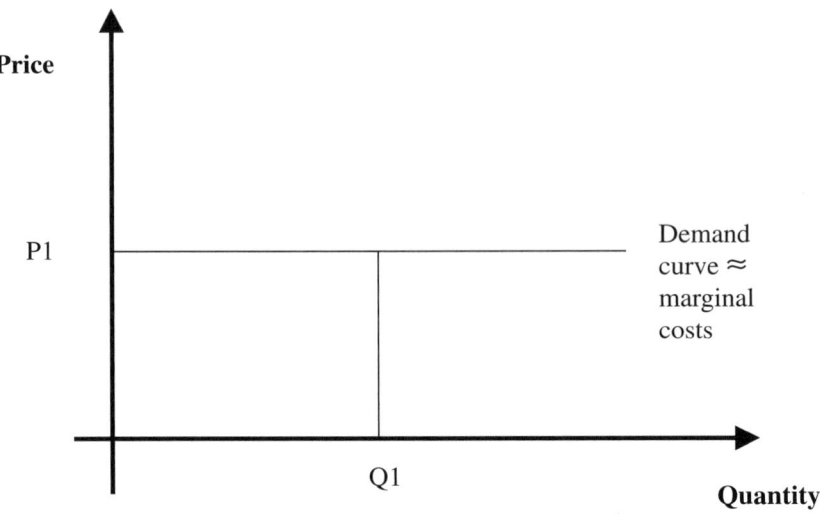

Figure 2.1   Perfect competition

At the other end of the scale, we find monopoly characterised by having only one company in the market, making the monopolist able to set both price and quantity (see Figure 2.2). Its demand curve is downward sloping. The monopolist will thus have to adjust his price according to his output – the more he wishes to sell the more he must lower the price.

A monopolist seeks to maximise his profits and will therefore price at the point that is beneficial for him. The monopolist's marginal revenue is the additional profit he gains from selling one extra unit. It is therefore less than the selling price. It means that the monopolist will not sell units where his marginal revenue is below his marginal costs (that is below point A in Figure 2.2). Based on this, the price the monopolist charges is Pm, which is higher than Pc, which is the price he would have charged if the market had been competitive. The monopolist's profit is the area of the trapezoid PmBAE. In a normal competitive market, the consumer surplus would be the FPcC triangle; however in a monopoly market this is decreased to FPmB. Besides the redistribution of welfare to the monopolist, there is also a deadweight loss to society illustrated by triangle BAC.

Where the monopolist sells the product to a distributor the monopoly profit will be passed on to the distributor as a cost. If he is also a monopolist, he will naturally set his price in the same manner as described above. The result is that when purchasing the product, the consumer in principle pays a price which contains two monopoly profits – in other words, two mark-ups or double

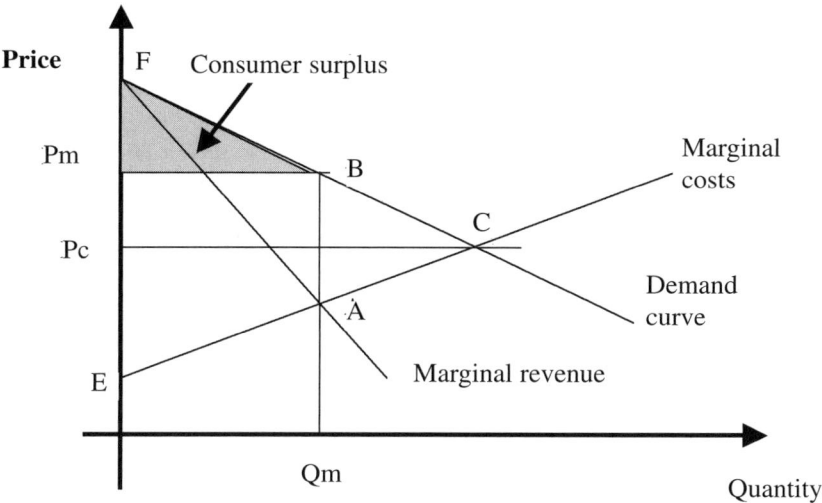

Figure 2.2   Monopoly

marginalisation.[103] The double marginalisation problem can be solved by vertical integration of the two monopolists, which would in principle eliminate one of the mark-ups. The double marginalisation problem shows that there is something worse than a monopoly – a chain of monopolies.[104] The double marginalisation problem can also occur in respect of complementary products. The monopolists will improve their gains if the products are sold in a bundle, although only if the products are complete complements, so that consumers buy both or none.[105] This was demonstrated by Cournot.[106] Two monopolists will price each complementary product at an individual profit-maximising price.[107] Were the two monopolists to integrate, they would sell the complementary products at a lower price than if they were not integrated, because the integrated monopolist can take into account the positive effect on the demand for one of the products of a reduction in price on the other. This is known as the Cournot effect.[108] This positive effect cannot be used by the two individual

---

103   Tirole, *supra* n. 93, 174–175.

104   Ibid., 175.

105   Nalebuff, *supra* n. 7, 37–38.

106   Cournot, Augustin, *Research into the Mathematical Principles of the Theory of Wealth* (Ed. N. Bacon, Macmillan, New York, 1897).

107   Tirole, *supra* n. 93, 174.

108   Ahlborn, Evans and Padilla (2004), *supra* n. 2, 322 and Kühn, Stillman and Caffarra, *supra* n. 11, 92.

producers. Tying can therefore eliminate the problem of double marginalisation. For instance in the case of franchising, tying will aid in the distribution of a franchisor's products, by the use of the franchisee as an alternative to vertical integration to promote the franchisor's own brand.[109]

> The reason is that each individual firm ignores the fact that a price cut would increase the demand for the complementary products of the other firms. This effect disappears when there is competition for each one of the complementary goods, such as the competition that would exist if single-product firms faced competition from a multi-product firm that sold its product separately.[110]

However, if single-product firms are faced with competition from a multi-product firm selling its product in a bundle, this will tend to make the Cournot effect reappear, as competition will be between the bundle and the single-firm complementary products.[111] The bundling creates a coordination problem for the consumer, and thus aggressive competition from the single-product firm towards the bundling company is not likely to occur.[112]

O'Donoghue and Padilla take a slightly different approach as they argue that the multi-product company is likely to have an increased incentive to compete aggressively on the tied market in spite of the bundle.[113]

In the *US Microsoft* cases, the Government had proposed a structural remedy for Microsoft's anti-competitive behaviour by splitting the company in two.[114] Although the remedy was accepted by the District Court, the Court of Appeals vacated the relief partly because it found the District Court had failed to offer a sufficient explanation for the relief sought.[115] From an economic perspective, such a break-up of a de facto monopoly would not have made sense, because the separate Microsoft firms would have maintained their monopoly status for their products and together they would have created double marginalisation.[116]

---

[109]   Hovenkamp, Janis and Lemley, *supra* n. 29, 21–10.
[110]   Kühn, Stillman and Caffarra, *supra* n. 11, 92.
[111]   Ibid., 93.
[112]   Ibid.; see also Carboja, De Meza and Seidman, *supra* n. 12, 296, who argue that this is evidence of strategic tying rather than price discrimination.
[113]   O'Donoghue and Padilla, *supra* n. 11, 483.
[114]   Evans, David, Nichols, Albert, and Schmalensee, Richard, 'United States v Microsoft: did consumers win?' [2005] *Journal of Competition Law and Economics* 497, p. 513 and *United States v Microsoft Corp.* Plaintiff's Proposed Final Judgment, Civil Action No. 98-1232 and 98-1233, 28 April 2000 (www.usdoj.gov/atr/cases/f219100/219106.htm).
[115]   *United States v Microsoft Corp.* (*US Microsoft III*) 253 F.3d. 34 at 46, 117, 119 and Evans, Nichols, and Schmalensee, *supra* n. 114, 513.
[116]   Gilbert, R.J. and Katz, M. L., 'An economist's guide to *U.S. v. Microsoft*' (2001) 15 *Journal of Economic Perspectives* 25, p. 41.

4.4. Concluding Remarks on Efficiency Reasons

The above-mentioned efficiencies are often recognised as valid explanations for tying and bundling as common business practices.[117] However, their significance is often overlooked or veiled in antitrust cases.[118] O'Donoghue and Padilla point to the standard argument that tying can be beneficial to some consumers, but the benefits should not restrict the separate sale of the individual components.[119] The presence of economies of scale and scope essentially means that savings can be made from joint manufacturing or distribution, and for this reason tying and pure bundling will occur in competitive markets even though some customers feel forced to accept the whole package.[120]

The commercial practice of tying, when analysed from an economic point of view, can have pro-competitive efficiencies, and thus tying should ideally be viewed as prohibited only when it has significant anti-competitive effects.

5. STRATEGIC REASONS FOR TYING

The section above focused on efficiencies that would mostly benefit both producer and consumer. However tying not only brings about efficiency gains, but it can also provide a purposefully sound strategy for gaining market shares and competing with rivals. 'When there is competition in the complementary products market, there are two reasons to cut price: market expansion and competition with rivals.'[121] It has been argued by some economists that tying is not a proxy for price discrimination, but rather a strategic game plan when the competition is imperfect in the tied market.[122]

5.1. The Leveraging Theory

The competition authorities fear tying mostly when it is applied by a dominant undertaking to expand its market power from one market into another.[123] Hence, the authorities tend to have been strict in their treatment of tie-ins despite their economic efficiencies. Tie-ins can exclude competitors as well as

---

[117] O'Donoghue and Padilla, *supra* n. 11, 483.
[118] Ibid.
[119] Ibid.
[120] Ibid., 483.
[121] Nalebuff, *supra* n. 7, 39.
[122] Carbajo, De Meza, and Seidman, *supra* n. 12, 283.
[123] Case T–201/04 *Microsoft Corp. v Commission*, [2007] ECR II-3601, paras 1351–1352.

limiting customers' ability to choose freely amongst the products available in the market.[124] Tying as a mechanism can increase market power, with the risk that the use of this market power in one market is applied to lever the exclusion of competitors and ultimately to foreclose a secondary market. According to the leverage theory, tie-ins are thought to be capable of driving other competitors out of the tied product market and thereby granting the tying undertaking the ability to make two monopoly profits.[125] The economic theories are therefore very much focused upon exploitative behaviour to increase profits. However, it should be highlighted that before the exploitation can take place exclusionary behaviour is essential to gain access to and later foreclose the tied product market.[126] 'Leveraging and foreclosure are examples of phenomena that arise because firms recognize the impact of current actions on the future competitiveness of the market.'[127]

Tying, though, is not the only form of restrictive conduct to which leverage theory applies. Refusal to supply or license while moving into a secondary market is conduct equally closely related to the leverage theory, and so are exclusive contracts and vertical mergers. Tying is often applied in conjunction with innovation, because tying is a strategic practice which can absorb some of the risk involved in introducing new products to the market. Innovations, often IP-protected in one way or another, can confer the ability to leverage, for instance via technological integration. Alternatively, they can threaten a company's monopoly.[128] It falls upon the competition authorities to determine whether the benefits of the innovation itself offset the anti-competitive effects of leveraging when leveraging implicates innovations.[129] In both the EU and US, the general position of the courts has been to condemn any practice which

---

[124]   Anderman, Steve, *EC Competition Law and Intellectual Property Rights, the Regulation of Innovation* (Oxford University Press, Oxford, 1998), p. 221. See also Nalebuff, *supra* n. 7, at 32, regarding consumer choice.

[125]   Jacobs, Michael, 'Third line forcing and tying arrangements – some comments on the United States' position', article published on the Blake Dawson Waldron web site: Competition and Consumer Protection (www.bdw.com.au/areas/tradespractices/tpa-new-4.htm (6 January 2004)).

[126]   Langer, Jurian, 'Bundling, A Four-Step Test to Assess the Exclusionary Effects of Bundling under Article 82 EC', Chapter 8 in Amato, Giuliano and Ehlermann, Claus-Dieter (eds), *EC Competition Law, A Critical Assessment* (Hart Publishing, Oxford, 2007), 298–299.

[127]   Kühn, Stillman and Caffarra, *supra* n. 11, 89.

[128]   Fisher, Franklin M., 'Innovation and Monopoly Leveraging' in Ellig, Jerry (ed.), *Dynamic Competition and Public Policy, Technology, Innovation and Antitrust Issues* (Cambridge University Press, New York, 2001), 138.

[129]   Ibid.

attempts to extend monopoly power.[130] However, it has long been a controversial topic among economists whether it is possible to extract two monopoly profits from leveraging and thereby a tying strategy.[131]

The traditional pre-Chicago School theory held that two monopolies would create two monopoly profits and therefore be more harmful than one monopoly.[132] 'If one monopoly is bad, surely two monopolies are worse.'[133] Scholars associated with the Chicago School have argued against such ideas, contending that when two products are combined and applied together there is only one product, and therefore only one monopoly to be exploited.[134] The full monopoly profits can be extracted from one of the products without excluding competitors in the market for the complementary product.[135] This theory is known as the single monopoly profit theorem or, as Bork calls it, the fallacy of double counting.[136] As Posner explains:

> A [fatal] weakness of the leverage theory is its inability to explain why the firm with a monopoly of one product would want to monopolize complementary products as well. It may seem obvious that two monopolies are better than one, but since the products are by hypothesis used in conjunction with one another . . ., it is not obvious at all. If the price of the tied product is higher than the purchaser would have had to pay in the open market, the difference will represent an increase in the price of the final product or service to him, and he will demand less of it, and will therefore buy less of the tying product.[137]

---

130   Cases 6 & 7/73 *Istituto Chemioterapico Italiano SpA & Commercial Solvents Corp. v Commission* [1974] ECR 223, [1974] 1 CMLR 309; Case 311/84 *Centre Belge d'Etudes du Marché-Télémarketing v Compagnie Luxembourgeoise de Télédiffusion SA and Information Publicité Benelux SA* [1985] ECR 3261, [1986] 2 CMLR 558; *Motion Picture Patents Co. v Universal Films Co.*, 243 U.S. 502, 518 (1917); *Aspen Skiing v Aspen Highlands Skiing Corp.* 472 U.S. 585 (1985); and Kaplow, Louis, 'Extension of monopoly power through leverage' (1985) 85 *Columbia Law Review* 516, p. 516.

131   Peritz, R.J.R., 'Competition Policy and Its Implications for Intellectual Property Rights in the United States', in Anderman, Steve (ed.), *The Interface between Intellectual Property Rights and Competition Policy* (Cambridge University Press, Cambridge, 2007), 188–189.

132   Feldman, *supra* n. 8, 2080.

133   Ibid.

134   Bowman, *supra* n. 74; Posner, Richard A., *Antitrust Law: An Economic Perspective* (University of Chicago Press, Chicago, Ill., 1976), and *Schor v Abbott Laboratories*.

135   Choi, Jay Pil and Stefanadis, Christodoulos, 'Tying, investment, and the dynamic leverage theory' (2001) 32 *RAND Journal of Economics* 52, 54.

136   Bork, *supra* n. 1, 372–373 and Kaplow, *supra* n. 130, 518.

137   Posner, *supra* n. 134, 171–172.

Ahlborn *et al.* argue that the single monopoly profit theorem merely holds that greater profit cannot be made through leveraging alone; the company must also create efficiencies via the tying arrangement, such as improvement of product quality or lower costs.[138] However, as Peritz notes, the leverage theory is dynamic, because the tying strategy is working over time attempting to achieve certain results in the future.[139] In comparison, the single monopoly profit theorem is a static theory, and thus the arguments raised above by Bork and Posner are true in a static short-run price theory model, but in a long-run model the consequences of tying may be more severe.[140]

The Chicago School argued that tying is beneficial to welfare overall and at no competition cost, therefore tying should be treated as *legal per se*.[141] Some of the welfare-enhancing effects have been highlighted above. These two arguments should be seen in conjunction with the policy of the US antitrust authorities not to treat price discrimination as illegal *per se*, because price discrimination has been named as the primary motivator for tying by the Chicago School.[142]

In comparison, the post-Chicago School theory identified certain circumstances where leveraging could indeed be harmful.[143] Whinston and Bishop and Walker are of the opinion that it is plausible to extract two monopoly profits unless the tied (complementary) products are applied in fixed proportions and the market for the tied product is competitive.[144] Ahlborn *et al.* note that the single monopoly profit theorem requires that the tied market be perfectly competitive; if this condition is not met the theorem may not be true.[145] This was proved by Whinston in his famous article 'Tying, Foreclosure, and Exclusion', in which he concluded that when a market contains economies of scale, strategic interaction and is not competitive, 'tying can make continued operation by a monopolist's tied market rival unprofitable by leading to the foreclosure of tied goods sales . . . such a strategy can be a profitable one for the monopolist, often precisely because of this exclusionary effect on market structure'.[146] Whinston's

---

[138]   Ahlborn, Evans and Padilla (2004), *supra* n. 2, 323.

[139]   Peritz (2007), *supra* n. 131, 188.

[140]   Ibid., 188–189.

[141]   Ahlborn, Evans and Padilla (2004), *supra* n. 2, 319.

[142]   Choi and Stefanadis, *supra* n. 135, 5 and Ahlborn, Evans and Padilla (2004), *supra* n. 2, footnote 119.

[143]   Feldman, *supra* n. 8, 2081.

[144]   See Whinston (1990) *supra* n 13; Whinston, M.D., 'Exclusivity and tying in *U.S. v Microsoft:* What we know, and don't know' (2001) 15 *Journal of Economic Perspectives* 63, p. 70 (Whinston 2001); Bishop and Walker, *supra* n. 22, para. 6.63 and Jacobs, *supra* n. 125.

[145]   Ahlborn, Evans and Padilla (2004), *supra* n. 2, 325.

[146]   Whinston (1990), *supra* n. 13, 855.

article lays the foundations for the post-Chicago theory in respect of tying.[147] His work has been developed and extended by scholars such as Carlton and Waldman, Nalebuff, Choi and Stefanadis.[148] Despite Whinston's theory being revolutionary, it has been found to have its own restrictions.

> The models demonstrate the theoretical possibility of tying to foreclose entry. They thereby provide a necessary correction to the Chicago view (among some adherents) that profitable foreclosure is never possible. The new models rest, however, on very restrictive assumptions. One is that bundling generates no efficiencies. Without allowing for possible efficiencies, the models cannot weigh the offsetting welfare effects of efficiency and foreclosure. Moreover, even if one were to incorporate efficiencies into the models, the remaining assumptions are so stylized that it is hard to know when they apply – if ever.[149]

The leveraging result thus depends upon the monopolist being committed and the tying leading to market foreclosure, otherwise the monopolist's tying strategy would be self-destructive, as it would instead lead to increased price competition in the market.[150] If the tied products are prefect complements and the tying creates foreclosure effects, the monopolisation of the tied product market may lead to lower prices in both markets, thereby reducing the incentive to tie.[151] In other words, the result of leveraging also depends upon the interrelationship of demand for the two products.[152]

Although both the Chicago and post-Chicago Schools have now been proved to have their limits, as their theories have been proved to be applicable only in certain specific circumstances, it did bring about the debate whether tying should remain illegal *per se* in the courts. The economic thinking did indeed manage greatly to influence the US courts as the judgments in *Jefferson Parish* and *US Microsoft III* reveal. In particular, the Chicago School opened the way for the thinking that tying strategies may not be purely anti-competitive but of an efficiency nature, and thus pro-competitive, by improving quality and decreasing costs.[153]

---

[147]   Hylton (2003), *supra* n. 36, 281, footnote 5.
[148]   Carlton, Dennis W. and Waldman, Michael 'The strategic use of tying to preserve and create market power in evolving industries' (2002) 33 *RAND Journal of Economics* 194; Nalebuff, *supra* n. 7; Choi and Stefanadis, *supra* n. 135.
[149]   Evans and Salinger, *supra* n. 27, 51.
[150]   See discussion further below; Ahlborn, Evans and Padilla (2004), *supra* n. 2, 326; and O'Donoghue and Padilla, *supra* n. 11, 486.
[151]   Ibid.
[152]   Ibid., 486.
[153]   *US Microsoft III*, at 84 and Ahlborn, Evans and Padilla (2004), *supra* n. 2, 319.

While leveraging is the most harmful strategy of tying, it is also a strategy that can be successful only for companies with market power; some even argue significant market power.[154] This conclusion should be kept in mind when assessing the legal approach to tying. As will be shown, leveraging has been viewed as the main cause of concern for the competition authorities.

5.1.1. Market foreclosure

The existence of tying will impede or deter a potential competitor from entering tied product markets. Hovemkamp *et al.* remind us that 'the real concern of foreclosure is not leverage but rather longer run stagnation and lack of competitive vitality'.[155] Equally, when tying leads to effective foreclosure, it denies consumers the choice of downstream products which would otherwise have been available.[156] This was the concern in *EC Microsoft*,[157] where the Commission feared that the tying of Windows Media Player (WMP) and Windows would foreclose competition and stifle innovation in the related media software encoding and management markets, because WMP would become the preferred choice for complementary content and application providers.[158]

Significant foreclosure on the tied product market can also lead to foreclosure or increased barriers to entry on the tying product market, because potential rivals are forced to enter both markets simultaneously rather than only one market, thereby increasing entry risks.[159] In high-technology industries, where the markets are often characterised by large sunk and fixed costs, network effects, occasional winner-takes-all product races, and thus few incumbents with market power, the risks involved in two-level market entries further increase.[160] Kühn *et al.* apply *EC Microsoft* as an example of how strong network effects in a two-sided market can provoke foreclosure. The media player market has two sides, the market for media players which run on PCs and the encoding software needed to run the content on the specific media player.[161] This side of the market is subject to strong network effects, because the more PCs that have the specific encoding format the more desirable it becomes for content providers to use that format.[162] Making multiple encod-

---

154 Ibid., 331.
155 Hovenkamp, Janis and Lemley, *supra* n. 29, 21–23.
156 Ibid.
157 Case COMP/C–3/37.792 *Microsoft*, [2005] 4 CMLR 965.
158 Ibid., para. 842.
159 Hovenkamp, Janis and Lemley, *supra* n. 29, 21–23.
160 Charles River Report, *supra* n. 14, 85.
161 Kühn, Stillman and Caffarra, *supra* n. 11, 101.
162 Ibid.

ing formats is costly, and therefore content providers tend to pick the format in which they can reach the most customers, rather than the best quality format.[163] This can create a snowball effect, tipping the market in favour of the software encoding manufacturer who is able to reach the most PCs.[164] Tying works as a trigger for such effect; however, market power in the operating system market is crucial.[165] Great value can be generated from more efficient and faster encoding methods to deliver content, in particular for internet-depended activities which rely on good delivery speed.[166] Therefore, there can be large negative effects of bundling strategies in these markets, and Kühn *et al.* are thus of the opinion that eliminating an alternative solution from the competition could reduce the quality of the software available, as the mere termination of investment in technology improvements is sufficient; it does not require actual foreclosure.[167]

5.1.2. Protection of the tying product market – defensive leveraging

It has been suggested by many scholars that the reasoning behind leveraging monopoly power from one market into another may not be a desire to gain market power in the adjacent market, but rather an attempt to deter entry into the primary market in order to protect the incumbent's market power in that market.[168] Hovenkamp *et al.* support the plausibility of this theory, but note that it is more feasible when the monopolies are consecutive rather than simultaneous.[169] If a monopolist produces product A, but a new product B is developed which is likely to take the place of A in the future, the monopolist thus risks losing his monopoly unless he strikes back. He can do so by tying his sale of the dominant product A with B.[170] This would force customers wishing to purchase product A also to purchase the monopolist's version of B. Hence, if B replaces A, the monopolist still has market power. Hovenkamp *et al.* give Microsoft's tie of its Windows operating system with its Internet Explorer browser as an example of protecting the monopoly in the primary market (this tie was the focus of *US Microsoft II and III*). Assuming that the browser develops into an operating system in its own right or works as a connector between multiple operating systems and thereby becomes a

---

[163] Ibid.
[164] Ibid., 101–102.
[165] Ibid.
[166] Ibid., 102.
[167] Ibid.
[168] See for instance Nalebuff, *supra* n. 7, 71; Ridyard, Derek, 'Tying and bundling – cause for complaint?' (2005) 26(6) *European Competition Law Review* 316, p. 317; and generally Feldman, *supra* n. 8, and Carlton and Waldman, *supra* n. 148.
[169] Hovenkamp, Janis and Lemley, *supra* n. 29, 21–21.
[170] Ibid.

competitor to Windows.[171] In such a scenario the tying arrangement may help Microsoft safeguard its dominant position in the operating systems market by allowing it to remain in control of the programme applied when a user moves from a Windows-based programme to an internet-based application.[172]

However, at any one time there is only one monopoly because the monopolies are successive, and thus only one monopoly profit; in other words, this theory corresponds better with that of the Chicago School. The anti-competitive effect of such defence strategy is a longer duration of the monopoly rather than increased prices for consumers.[173] However, one can question whether there is not also a risk of innovation stagnation if such a strategy is successful. In relation to the recent *EC Microsoft* case, it has been argued that Real Networks is a possible crucial component for the development of a middleware which could pose a serious threat to Microsoft's monopoly on the PC operating system market, much like Netscape's Navigator in the past.[174]

Actually, Kühn *et al.* argue that the software industry is one of the few industries in which it is possible to detect a link between today's bundling and tomorrow's competition, as in the software industry the application programme interfaces to which application developers write their software are not standardised, and thus the application developers are forced to write either to several APIs – a rather costly affair – or merely for the most widely distributed system,[175] the second being the more profitable and therefore the strategy most often applied.[176] The more applications are written for one system, the more this will enhance the value of the system to consumers and thus create a self-reinforcing network effect.[177] This is known as the applications network effect and seen as the main reason for a market to tip towards monopoly, because it leaves a lasting effect upon it and allows the monopolist to produce a 'sunk cost effect' on consumers through the tying.[178]

In comparison, Carlton and Waldman's theory is based on a monopolist producing two products where one – the tying product – can be applied on its own and the other – the tied product – is a complementary product and can be used only in combination with the tying product, for instance, a computer and a mouse.[179] The entry deterrent would be the fact that a potential competitor

---

[171]   Ibid., compare with Case T–201/04 *EC Microsoft*, para. 911.
[172]   Ibid.
[173]   Ibid.
[174]   Ridyard, *supra* n. 168, 317 and Case T–201/04 *EC Microsoft*, para. 911.
[175]   Kühn, Stillman and Caffarra, *supra* n. 11, 98–99.
[176]   Ibid.
[177]   Ibid.
[178]   Ibid., 98–100.
[179]   Carlton and Waldman (2002), *supra* n. 148, 198.

on the complementary product market would be competing against the monopolist.[180] In order for the theory to work, however, there must be high costs involved in entering the complementary product market or the potential competitor would enter this market as well as the primary product market and the strategy of foreclosure would be suppressed.[181] If the costs of entering the complementary product market are low or non-existent, the theory will be successful only if there are network externalities present in the market.[182] This form of strategy also works as a device to deny the entrant network effects or economies of scale.[183] This strategy seems to be a popular reason behind tying. For example, if Microsoft could reduce Real Networks' market share by including its WMP in the Windows operating system, this would ensure that fewer companies would design programs for RealPlayer than for WMP.[184] In extreme situations tying can lead to the elimination of an adjacent market and is hence in principle a form of vertical integration, which increases the barrier to entry for potential competitors as well as raising their costs.[185]

Nalebuff suggests that tying products together can only help the monopolist fend off potential entrants: once the entry takes place an incumbent would prefer not to apply tying as a strategy.[186] In order for the incumbent to succeed with such a strategy it must be fully committed to the tying, and hence sell its products only in a bundle. The incumbent can thereby use the value of one product to cross-subsidise the other product where entry is attempted, rebuff the entrant any sale, and protect its monopoly position.[187]

Applying the theory of defensive leveraging to *EC Microsoft* the real harm to consumers is not the current or future lack of choice in media players, but the restriction of potential benefits that could develop from Microsoft having competition in the operating systems market.[188] There is no doubt that the defensive leverage theory fits well with Microsoft's recent actions.

5.1.3. Undercutting rivals' price by tying
As shown above, tying can eliminate double marginalisation where it is practised by a monopolist by producing two complementary products. Such a strategy is efficient for the monopolist and undercuts competitors' prices. In

---

[180]   Ibid.
[181]   Ibid. and Ahlborn, Evans and Padilla (2004), *supra* n. 2, 327–328.
[182]   Carlton and Waldman, *supra* n. 148, 206–207; Ahlborn, Evans and Padilla (2004), *supra* n. 2, 328; and Whinston (2001), *supra* n. 144, 72.
[183]   Nalebuff, *supra* n. 7, 55.
[184]   Ibid., 56 and Case T–201/04 *EC Microsoft*, para. 911.
[185]   Nalebuff, *supra* n. 7, 56.
[186]   Ibid., 59.
[187]   Ibid., 58–59.
[188]   Ridyard, *supra* n. 168, 318.

response to such a strategy, it could be expected that the competitors would lower their prices too, even though they were selling only one of the products or both individually.[189] Because of the elimination of double marginalisation, bundled pricing will often result in price reductions.[190] That said, as single-product competitors are also likely to lower their prices, the rise in profits and market shares can be uncertain and unclear.[191]

According to Whinston, tying will still deter entry by potential competitors into the tied product market as they know that the incumbent company will lower prices if they enter the market to make entry unprofitable for them. This naturally enables the incumbent company to maintain sales in the market.[192] However, like the Chicago School theory, Whinston's theory is not robust because, if there are alterations in the underlying assumptions, the outcome will change radically. Whinston's theory requires that the monopolist is able to commit to tying and that the tying leads to the foreclosure of the market; if not the strategy would be unsuccessful and increase competition in the market.[193]

There is another side to the coin as well – the tying arrangement can increase the rivals' costs as it can require a competitor to enter both markets to compete effectively with the tying company. This naturally raises the barrier to entry to the markets, and the risk of entering two markets is greater than for one.[194] This is a crucial point, as it means that, even if the Chicago School economists are correct in their theory of tying conduct not being able to lever market power, tying can still raise rivals' costs and thereby be a profitable strategy.[195]

5.2. Competition Mitigation

Once entry by a competitor has taken place, tying can be employed by the incumbents to take the edge off the competition and create product differentiation.[196] Carbajo *et al.* have demonstrated that bundling can create a strategic incentive and be profitable because it can induce rival companies to compete less aggressively, even in the presence of imperfect competition.[197] If

---

[189] Nalebuff, *supra* n. 7, 39.

[190] Ibid.

[191] Ibid.

[192] Whinston (1990), *supra* n. 13, 855.

[193] Ahlborn, Evans and Padilla (2004), *supra* n. 2, 326.

[194] Baldwin, W.D. and McFarland, D., 'Tying arrangements in law and economics' (1963) 8: 5–6 *Antitrust Bulletin* 743, pp. 773–775.

[195] Martin, S., *Industrial Economics, Economic Analysis and Public Policy* (2nd edn, Prentice Hall, Englewood Cliffs, NJ, 1993), 442.

[196] Nalebuff, *supra* n. 7, 45–46.

[197] Carboja, De Meza and Seidman, *supra* n. 12, 283 and 285.

company A produces two products, X and Y, and company B produces only Y, and if company A sells X and Y untied, then there will be price competition between A and B in the Y product market. However, if A sells X and Y only in a bundle, high value customers will purchase the bundle and the rest will content themselves with just Y.[198]

Chen adopts a different approach, albeit with the same result, as he assumes that the primary market (A) is a duopoly compared to the norm of analysing tying behaviour with the primary market as a monopoly.[199] Moreover, he assumes that there is perfect competition in the secondary market (B). His justification for these assumptions is that observations have shown that many companies applying tying as a strategy are indeed active in such markets and companies apply the tie as a tacit collusion strategy to differentiate between products and thereby take the pressure away from price competition.[200] Companies 1 and 2 are both present in market A but, as mentioned, market B is competitive. Chen's theory is that the two companies would divide market A between them, and hence Company 1 would sell only A whereas Company 2 would sell a bundle of A and B. As Chen notes, bundling, because of its product-differentiating ability, becomes an equilibrium strategy of competing companies.[201] Mixed bundling is the common strategy in competitive markets, whereas pure bundling prevails under a monopoly.[202] Pure bundling will result in higher profits for all companies in a market, and thus this will be the preferred option if it is possible for the companies to commit to the bundling, though the social surplus suffers as a result.[203]

In conclusion, tying is not being used here as an attempt to exclude competitors from another market. Instead, it is applied to divide the tying product market through a sophisticated way of creating product differentiation.[204] It is therefore more closely related to defensive than ordinary leveraging.

5.3. Gaining Competitive Advantage

It is in some ways reiteration to include a section on tying to gain competitive advantage, as it must be assumed that this is the overall purpose of tying. When companies apply tying as a strategy, they expect to gain some kind of

---

[198] Ibid.
[199] Chen, Yongmin, 'Equilibrium product bundling' (1997) 70 *Journal of Business* 85, 85–86.
[200] Ibid.
[201] Ibid., 99.
[202] Ibid.; compare though with Section 6.
[203] Ibid.
[204] Nalebuff, *supra* n. 7, 46.

competitive advantage over their rivals, whether this is in the shape of cost reductions or price discrimination. However, the focus here is on bundling or tying when applied as part of a network or in cooperation with other companies.

5.3.1.   Creating network externalities

The more products that are included in a bundle the more attractive the bundle becomes for the consumer, in comparison to purchasing each product individually.[205] This is why we find that companies create a bundle or alliance with other companies. Good examples are offered by the cable television industry's package subscription to particular channels, phone and internet connection, and shoppers' loyalty cards and clubs like Nectar and Airmiles. As Nalebuff illustrates with the *Aspen skiing* case, the different bundles or alliances will then compete against each other, but once the customer has bought into these bundles or alliances the companies within the alliance will have to compete for the customer's business.[206] Being part of the alliance can prove vital to a company, and hence the exclusion of companies from an alliance will lead to loss by the excluded company, which will be much higher than the loss to the alliance.[207]

In particular, in high-technology markets there are tendencies towards demand-side network effects, which can lead to the market tipping towards one single technology or company.[208] In markets where winner takes all aggressive strategies are likely to pay off.[209] In these situations, tying or bundling can trigger the network effects or enhance them, if applied at the right time; they can be instrumental in tipping the market in one particular direction.[210]

EC Microsoft is an interesting example of this, as the media player market is very close to this tipping point, and the Commission's interference was partly an attempt to prevent the tipping of the market.[211] However, tipping or snowballing is only achievable when there is some sort of bottleneck restricting the industry.[212] In the case of the media players the restriction is the fact

---

[205]   Katz, Michael L. and Shapiro, Carl, 'Network Externalities, Competition, and Compatibility' (1985) 75 *American Economic Review* 424, p. 424.

[206]   Nalebuff, *supra* n. 7, 49–53.

[207]   Ibid., 53.

[208]   Lind and Muysert, *supra* n. 37, 89.

[209]   Ibid.

[210]   Ibid.

[211]   Hellström, Per, presentation given at 'The Microsoft Judgment' Conference, 25 September 2007, BIICL, 17 Russell Square, London and *EC Microsoft* Commission Decision, para. 842.

[212]   Veljanovski, C., 'EC antitrust in the new economy: is the European EC Commission's view of the network economy right?' (2001) 22 *European Competition Law Review* 115, p. 116.

that music produced to be played on one type of media player cannot be played on another media player. Microsoft's business strategy to promote its media player with its operating system makes it more attractive to both customer groups: the content providers and consumers. The only way of breaking the network is by developing a new generation of the product or a similar product with superior technology. Even so this can be difficult because it requires the majority of the users of the network to switch in order for the switch to be beneficial for the individual user.[213] An example of this is the switch from vinyl records to CDs or from video tapes to DVDs to Blu-ray-Disc.

Nevertheless intervention from competition authorities by breaking up the network can cause more harm than good to consumers; as stated above, the larger the network the more beneficial for consumers.[214]

5.4. Price Obfuscation

Prices become obscure because consumers cannot always see through the relationship between price and cost of the bundled package and that of the individual products in the bundle.[215] The introduction of Microsoft's PowerPoint into its Office package illustrates price obfuscation. PowerPoint was at the time not an industry standard, but Microsoft added it to its Office package, while setting a high stand-alone price for it.[216] The consumers' perception was therefore that they were getting a high-value product within the package deal, but naturally, the real discount remains hidden.[217] Adding a product or component to a bundle is therefore a good introduction strategy for new products, as market price is difficult to assess without the bundle.[218]

Price obscuring does not appear to be an effective strategy to distort competition, but rather affects consumer choice. As Nalebuff has demonstrated, price obscuring mainly occurs in competitive markets such as foreign package holidays and insurances.[219] When tying is applied to obscure price, it is an efficient tool for entering a market, although there is a risk that it will mislead consumers and may therefore be illegal under consumer protection legislation.

---

[213] Ibid., 115.
[214] Ibid., 116 and Ahlborn, C., Evans, D.S. and Padilla, A.J., 'Competition policy in the new economy: Is European Competition Law Up to the Challenge?' (2001) *European Competition Law Review* 156, p. 159.
[215] Nalebuff, *supra* n. 7, 61.
[216] Ibid., 62.
[217] Ibid.
[218] Ibid.
[219] Nalebuff, Barry and Majerus, David, 'Bundling, tying, and portfolio effects' DTI Economics Paper No 1, Part 2 – Case Studies, February 2003, p. 105.

As will be seen from the following chapters, price obscuring has not caused major anti-competitive problems in case law, but seems to be a strategy applied in competitive industries.

5.5. Technological Integration

Examples of technological integration can be detected in many different types of markets, such as cars sold with engine and steering wheel, and mobile phones with cameras. The above examples of technological integration will never fall within antitrust concern as they appear in competitive markets rather than monopolies.[220] Using *US Microsoft III* as an example, Evans highlights that there clearly must be benefits from product integration, since all or most other companies also integrate the same components.[221] Other economists have agreed with this view, highlighting that product integration is a natural progression in high technology industries such as the software industry, acting as an incentive to innovate as well as benefiting consumers and potentially improving the functioning of the bundle.[222] Interestingly, in the car industry, which was once seen as a high technology industry but which now has grown into a more mature industry, motor vehicle manufacturers still eagerly apply various forms of tying in respect of optional equipment. For instance, a car stereo used to be an additional feature and offered separately from the car; then it became an optional feature with a choice of several different types; and finally it is now sold with the car as standard.[223]

However, if one company succeeds in creating strong network effects, it can affect the other competitors' ability also to benefit from technological integration, as the risk of entering the other market, be it either the tying or tied product market, may be too great.

Whinston demonstrated that in certain circumstances a monopolist could leverage monopoly power through commitment to technological integration.[224] However, in his examples it was assumed that the technological integration was incompatible with competitors' products in the tied product market.[225] If this is not the case, and consumers are able to use competitors'

---

[220] Kühn, Stillman and Caffarra, *supra* n. 11, 88.

[221] Evans, D.S., 'All the facts that fit: square pegs and round holes in US v. Microsoft' (1999) 22 Regulation no. 4 p. 1–10 (www.cato.org/pubs/regulation/regv22n4/evans.pdf); Kühn, Stillman and Caffarra agree with this view: Kühn, Stillman and Caffarra, *supra* n. 11, 114.

[222] Evans, Padilla and Polo, *supra* n. 59, 509 and 514. See also Whinston (2001), *supra* n. 144, 70.

[223] Evans and Salinger, *supra* n. 27, 75–76.

[224] Whinston (1990), *supra* n. 13.

[225] Whinston (2001), *supra* n. 144, 72.

products together with the bundle, the question is whether the benefit of competitors' products outweighs the marginal costs of the bundled products.[226] It is in principle this question which the courts on both sides of the Atlantic had to assess in the *Microsoft* cases, because only 'incompatibility or some other factor that makes it infeasible or unattractive to add the rival's version of [the tied product] onto the system is necessary for a physical tie to have an exclusionary effect'.[227]

Nalebuff agrees and argues it is possible for a monopolist to control adjacent markets by physical product integration, changing interface information, withholding or delaying the information regarding the change or simply making the product incompatible.[228] However, it is not necessarily in the monopolist's interest to exclude companies in adjacent markets, as consumers may not continue to purchase the product if certain other products cannot be applied in conjunction with it.[229] For instance a restriction on compatibility with a well-established complement could reduce the potential *ex post* value of the monopolist's product system.[230] Although this may be true for some adjacent markets it is not for others, because the company can still benefit from preventing other competitors of the complementary products from entering the market, or even exclude such competitors from the market, especially those causing significant threats to the company's monopoly in the future.[231] The adjacent market is easiest to control if it is competitive; otherwise the preferred and well-established complement producer has an opportunity to gain some of the profit.[232] Nalebuff concludes, '[technological integration] is a symptom rather than a cause. A firm facing competitors will not want to engage in . . . [technological integration] unless it truly leads to product enhancements that outweigh the loss in diminished complements . . . .'[233] Firstly, it is unclear exactly what Nalebuff means by describing technological integration as a 'symptom'. He could mean that technological integration is something companies will engage in as a result of market power and wanting to lever into a secondary market; in other words it is a tool for leverage, or it is a tool for greater efficiency and meeting consumer demands for a more user-friendly product. This corresponds well with the findings above as regards search costs. Nalebuff does not identify conditions or evidence that will establish whether

---

[226]   Ibid.
[227]   Ibid., 72.
[228]   Nalebuff, *supra* n. 7, 57.
[229]   Ibid.
[230]   Ibid.
[231]   Ibid.
[232]   Ibid.
[233]   Ibid., 58.

the company is pursuing either. He indicates that harmful technological integration is a consequence of market power. This would explain the term 'symptom', but if this is the case, he creates a distinction between harmful and beneficial technological integration. The former takes place when the company holds market power. There is nothing special or new in this conclusion. Nalebuff is thus merely noting that companies holding market power would be likely to engage in technological integration to foreclose or exclude other competitors, whereas companies facing competition would engage in technological integration only if it led to improvements in products which outweighed losses of sales of complements. Such a distinction may be one step too far, as it would mean that a company facing competition would not consider technological integration even if this was what other competitors were doing, if the improvement to its product was less than the loss of sale of complements, and a dominant company would only engage in technological integration to exclude rivals, not to improve products or reduce its own costs. This is not the case, since the efficiencies and benefits of technical bundling often outweigh its disadvantages and thus, regardless of losses to sale of complements, companies are likely to engage in technological integration, especially if other competitors are also doing it.

Nalebuff continues '. . . [the] law can offer some protection and this may eliminate the more egregious behaviour, but we should not expect that enforcement to be easy'.[234] Neither the US antitrust nor EC competition legislation covers technological integration; only case law has indicated that technological integration is illegal, especially when the company holds market power. Instead of introducing a test for distinguishing between harmful and efficient technological integration, all Nalebuff seems to say is that technological integration performed by a company with market power is more likely to be harmful than when it is performed by companies facing competition. That said, the law and its upholders (the courts and the competition authorities) may not be best placed to assess whether a technological integration is a true product-enhancing integration or merely an attempt by a company to exclude competitors.[235]

Ahlborn *et al.* suggest that tying is divided into classes, making a distinction between contractual tying and technological integration.[236] They do so with US antitrust law in mind, albeit that their suggestion can easily be adapted to EC competition law. Although they recognise that a more relaxed approach should be adopted towards tying, they favour a moderate *per se legality* approach to technological integration, whereas they favour a *rule of*

234   Ibid.
235   See *US Microsoft II* at 950, fn 13.
236   Ahlborn, Evans and Padilla (2004), *supra* n. 2, 339–340.

reason approach to contractual tying where the company holds significant market power.[237] In comparison, the US courts have treated technological integration with a moderate *per se* illegal approach, which actually has been more lenient than the courts' illegal *per se* treatment of contractual tying.[238] Moreover, Whinston highlights that physical tying is ubiquitous because it creates value and even a production process can be viewed as a process of physically tying.[239] This is very much in line with Bork's view that all products can be divided into smaller components – physical tying or technological integration is thus ever-present.[240] Whinston therefore questions the reasons for contractual tying, as he recognises that these are less well understood.[241]

Although technological integration is seen as a strategic reason for tying it clearly offers benefits in the shape of innovation. There is therefore good reason to listen to the economists who call for a more lenient approach to this form of tying in comparison to contractual tying.

5.6. Tying as an Incentive to Innovate

Building upon Whinston's model, Choi applies a three-stage model where companies engage in R&D competition prior to the pricing game in order to prove that tying can have additional advantages to the tying firm in the R&D market and that tying strategies can hinder access to consumers by rival companies.[242] In the first stage of Choi's model, Company 1 decides whether to bundle. In the second stage, the two companies adopt cost-reducing R&D activities, and in the third stage a price game develops with the cost structure inherited from the R&D achievement.[243] Choi finds that tying can indicate a commitment towards more aggressive R&D investment.[244] This finding supports the argument that tying by a dominant company can

---

237   Ahlborn, Evans and Padilla (2004), *supra* n. 2, 339–340.
238   See cases such as *Telex Corp. v International Business Machines Corp.*, 367 F. Supp 258, 268 (N.D. Okla. 1973), rev'd, 510 F.2d 894 (10th Cir. 1975), *ILC Peripherals Leasing Corp. v International Business Machines Corp.*, 448 F. Supp. 228, 233 (N.D. Cal. 1978) and *ILC Peripherals Leasing Corp. v International Business Machines Corp.*, 458 F. Supp. 223, 239 (N.D. Cal. 1978). Compare with *Article 82 Guidance*, *supra* n. 54, 17 and US Department of Justice, 'Competition and Monopoly: Single-Firm Conduct Under Section 2 of the Sherman Act' (2008) (www.usdoj.gov/atr/public/reports/236681.htm) (*Section 2 Report*), 88–89.
239   Whinston (2001), *supra* n. 144, 70.
240   Bork, *supra* n. 1, 378–379.
241   Whinston (2001), *supra* n. 144, 70–72.
242   Choi, *supra* n. 69, 22.
243   Ibid.
244   Ibid., 17.

suppress innovation on the tied product market, because foreclosure of the product market converts into foreclosure of the R&D market.[245] 'The change in R&D incentives through bundling enables the tying firm to capture a larger share of dynamic rents. If this effect outweighs the negative effect of more aggressive price competition, bundling will be a privately optimal strategy even in the absence of exit by the rival firm.'[246]

5.7. Concluding Remarks on Strategic Reasons

This section pinpointed several strategic reasons behind tying. However, there is one common and obvious denominator to all these reasons – they will in severe cases result in market foreclosure. There is nothing to stop a company from reaping the benefits from several of these strategic reasons at the same time; for instance, if the company engages in technological integration, it can reap the rewards the bundling creates as an entry deterrent as well as the benefits it will create in terms of competitive advantage, price obfuscation and innovation. Nalebuff noted that as regards technological integration, it is a symptom rather than a cause. This conclusion can be extended to many of the strategic reasons dealt with above. They are strategies which permit the companies to be more efficient, although at times the strategies will ultimately result in market foreclosure.

This section discussed the different ways in which tying can be applied to achieve different aims. Some will be pursued only when specific market conditions are present and, more importantly, most are harmful only when undertaken by a company holding significant market power.

6.   EMPIRICAL EVIDENCE

As can be seen from the above analysis, there has been no shortage of theoretical contributions to the justifications for tying and bundling. However, there has been limited empirical evidence.

Evans and Salinger have produced some empirical evidence, focusing their analysis on three types of products: cold medicines, foreign electrical adaptors and optional vehicle equipment.[247] The assessment of these three competitive markets is interesting, not only for the outcome of Evans and Salinger's analysis, but also because it mirrors part of Areeda *et al.*'s treatise in which they

---

[245]   Ibid.
[246]   Ibid.
[247]   Evans and Salinger, *supra* n. 27.

propose a test for establishing whether products are separate.[248] The test consists of, amongst other things, an assessment of several rationales. The first and most important rationale of the Areeda *et al.* analysis is Competitive Market Practices.[249] The test assesses whether companies in a competitive environment tie the products in the same manner as the defendant. If so it indirectly offers evidence that the two products are in fact two components of one product.[250] However, in the empirical evidence offered by Evans and Salinger there are no dominant companies, in which tying conduct can be compared with the action of competitive companies; on the other hand, the empirical evidence collected proves that tying and bundling are common business practices.

The cold medicines market has been identified as one of the most competitive over-the-counter drug markets in North America, where producers offer an array of overlapping cold and cough remedies.[251]

> Someone who has a cold (but not a cough) and would like a remedy to help get through the day typically has two needs: a decongestant (to relieve stuffiness) and a pain reliever (to relieve headaches). Someone who has a headache but no congestion does not need a decongestant; someone who has congestion but no headache has no need for a pain reliever – there are thus distinct demand groups for each of the components.[252]

Evans and Salinger found that significant discounts were given for the bundled products, although mixed bundling was the prevalent practice.[253] They argued that the most likely explanation for this was marginal costs savings, although price discrimination could not be completely ignored.[254]

The market for foreign electrical adapters is prone to tying, and thus most customers are forced to purchase products they do not want.[255] That said, the market is highly competitive and has low barriers to entry.[256] Evans and Salinger found that the explanation of the tying in this market was economies of scale (in relation to packaging and shelf space). They rejected the possibility of price discrimination and leveraging mainly due to the competitiveness

---

[248]   Areeda, P., Hovenkamp, H. and Elhauge, E., *Antitrust Law, An Analysis of Antitrust principles and their Application* (Little Brown & Company, New York, 1996), vol. X at para. 1744.
[249]   Ibid., at para. 1744, p. 197.
[250]   Ibid., at para. 1744d, p. 202.
[251]   Evans and Salinger, *supra* n. 27, 66.
[252]   Ibid.
[253]   Ibid., 70–71.
[254]   Ibid., 70.
[255]   Ibid., 72.
[256]   Ibid.

of the market.[257] However, they noted, 'with significant market power it may be hard to distinguish efficiency from less innocent theories'.[258] It will always be difficult to assess the incentives behind tying strategies, and thus certain scepticism must be maintained in markets dominated by one or two major companies. The findings of Evans and Salinger also emphasise the importance of identifying the right form of evidence and the defendant being faced with a realistic standard of proof to be able to demonstrate sufficiently accurately that a chosen tying strategy is valid on efficiency grounds.

Finally, in the market for optional equipment for vehicles, Evans and Salinger found an interesting trend, namely that choice paves the way for mixed bundling and mixed bundling paves the way for tying. In other words, optional equipment becomes standard over time.[259] In the US, the market for vehicles was for many decades oligopolistic, but during the 1970s this changed, and in 2003, when this study was undertaken, the market was much more competitive and prices had fallen significantly.[260] The trend in the market towards more tying, while the market had at the same time become more competitive, eliminated the possibility of price discrimination as a motive for the tie. Moreover leveraging can equally be excluded as a motive because most of the tying trends were initiated by entrants into the market, not the incumbents.[261] Tying basically occurred in order to cut costs in otherwise complex production and distribution.[262]

All three examples illustrate some important points regarding consumer choice. Elimination of consumer choice ranks high on the competition authorities' list of reasons for tying being harmful. In the example of the cold medicine market, the offer of bundled products ensures a vast variety of products and choice, but also, from a practical perspective, limits the number of pills consumers have to take per illness.

In the foreign electronic adapters markets, consumers are forced to take the bundle despite a competitive market with low barriers to entry (that is the closest a real world market will get to a perfect competition situation), and thus consumer choice is very restricted. This finding is at odds with Areeda *et al.*'s theory of all companies bundling in a perfectly competitive market, because of profitability and consumer preference.[263] However, in the foreign electronic adapters market, the only reason for the bundling seems to be costs efficien-

---

257   Ibid., 73–75.
258   Ibid., 72.
259   Ibid., 75.
260   Ibid., 76–77.
261   Ibid., 82.
262   Ibid., 83.
263   Areeda, Hovenkamp and Elhauge, *supra* n. 248, at para. 1744, p. 197.

cies for producers, a factor to which Areeda *et al.* have not given much consideration in their theory, indicating that the theory is either wrong or very sensitive to small changes in market conditions, rendering it in principle useless.[264] Interestingly, the empirical evidence was gathered in 2003; now, in 2009, foreign electronic adapters are no longer just tied together; rather, integrated single adapters which can be applied to several electrical systems worldwide are now available.

In the market for optional equipment for vehicles tying is the norm. Although this may relate to the complex production of vehicles rather than consumer demand, there is evidence that too much choice for consumers can run the risk of them not making a decision to purchase.[265] Thus in respect of complex products such as vehicles, consumers are better off with a restricted choice. This also corresponds well with the findings above regarding search costs, where product integration seemed to be partly a response to consumer preferences.[266] Finally, over time optional equipment becomes standard; therefore consumers expect the vehicle to contain certain equipment rather than being confronted with a choice of having it or not.

Interestingly, from a consumer perspective, the tying in the foreign electronic adaptor market (in 2003) is therefore the most harmful for consumers, as the product is relatively simple yet tied, and that despite the market being competitive. Less so now, where the market has moved to technological integration, equivalent to the optional equipment for vehicles market.

Although the empirical evidence sampled by Evans and Salinger focused on competitive markets with no company holding a dominant position, it identified two important issues. First, it confirmed that tying is a business strategy also undertaken in competitive markets, and thus leveraging is (often) not the main aim.[267] The reduction of fixed costs seemed to play an important part for companies when deciding to tie or bundle. Interestingly, this incentive is equally important for monopolists, and thus worth taking into consideration in future tying cases.[268] Second, the empirical evidence illustrated that tying is a complex issue and thus requires a more refined test by the authorities before it can be identified as mainly pro- or anti-competitive. In particular, the identification of the efficiencies and the motives behind the tying strategy require

---

[264]   Compare with Kühn, Stillman and Caffarra, *supra* n. 11, 107 and Section 4.2.2 above, 'Reduction in search costs'.

[265]   Nalebuff, *supra* n. 7, 32 and Dhar, R. and Nowles, S.M., 'To buy or not to buy: response mode effects on consumer choice' (2004) 41 *Journal of Marketing Research* 423, 431–432.

[266]   Section 4.2.2 above.

[267]   Evans and Salinger, *supra* n. 27, 85.

[268]   Ibid.

more attention in comparison to what is now offered in the current legal tests applied by the courts on both sides of the Atlantic. Moreover, the empirical evidence indicates that applying just one test for assessing both anti-competitive effects of tying and restrictions on consumer choice is insuffi-cient, as consumer choice may not always be harmed although the tying is anti-competitive. Equally, consumer choice can be harmed even if the tying is conducted within a competitive market.

7. HOW SHOULD TIE-INS BE TREATED? – AN ECONOMIC PERSPECTIVE

In the above analysis, several incentives have been identified for tying and bundling. However, no test as such has been suggested by economists to iden-tify the purpose behind each individual tie, and the empirical evidence assessed above illustrates that reasons behind tying in specific markets can be identified but they are difficult to verify. There is even uncertainty whether tying or bundling really is as harmful as the courts seem to think. Whinston notes:

> [What] is striking about the area of . . . tying . . . is how little the current literature tells us about what [the effects of tying] are likely to be. This state of (non) knowl-edge is . . . responsible to a significant degree for the very strong but differing beliefs that economists often have about whether . . . tying [is] likely to have welfare-reducing anticompetitive effects.[269]

As highlighted by Padilla and others,[270] three essential conclusions can be drawn based on the economic assessment of tying. First, tying is a common business practice in many industries arising from efficiency incentives.[271] Second, Whinston's and Carbajo *et al.*'s theories, for instance, illustrate how difficult they are to substantiate and how susceptible the models are to minute changes in the markets assessed, making it difficult positively to identify anti-competitive effects.[272] Third, as Evans and Salinger also demonstrate in their empirical evidence investigation, it is difficult to weigh potential efficiencies

---

[269]   Whinston (2001), *supra* n. 144, 79.
[270]   See Ahlborn, Evans and Padilla (2004), *supra* n. 2, 339–340 and O'Donoghue and Padilla, *supra* n. 11, 491.
[271]   As emphasised by Evans and Salinger's empirical evidence: Evans and Salinger, *supra* n. 27.
[272]   Whinston 2001, *supra* n. 144, 79; Carboja, De Meza and Seidman, *supra* n. 12; and O'Donoghue and Padilla, *supra* n. 11, 491.

against potential anti-competitive harm.[273] Many economists therefore advocate that the original approach of illegalising tying and bundling *per se* is not justified.[274] Instead, they propose adopting a *rule of reason* analysis, especially in relation to high-technology markets.[275] Ridyard notes, 'the fact that bundling is ubiquitous underlines . . . [that] there can be no economic sense in a *per se* rule "against" bundling'.[276] Lind and Muysert support this argument by referring to the recent decision by the US Court of Appeals in *US Microsoft III*[277] to send the case back to the lower court for a *rule of reason* analysis of the alleged abuse of tying.[278] They have identified the following issues which need to be examined as potential justifications for tying or bundling in dynamic industries:

(1) the potential for price discrimination to lead to efficient recovery of fixed costs;
(2) the need for firms selling complex systems to protect their reputations;
(3) the ability of bundling to reduce prices and increase sales;
(4) the potential for cost savings from bundling; and
(5) rational product integration.[279]

However, these points have to be counterweighed by the risks of market foreclosure and the discouragement of companies from entering a market, and hence a reduction in innovation of new products.[280]

> Firms competing in markets involving large fixed (and sunk) costs face significant risk when entering markets, and require reasonable prices and volumes to survive. Commercial practices that reduce the share of the market available for the entrant to contest, reduce the price the entrant can expect for its products or substantially raise the risk, may have considerable potential to deter entry, and reduce innovative investment in new products.[281]

---

[273] Evans and Salinger, *supra* n. 27, 72.
[274] See ibid.; Ahlborn, Evans and Padilla (2004) *supra* n. 2; Hylton and Salinger, *supra* n. 4; Lind and Muysert, *supra* n. 37; O'Donoghue and Padilla, *supra* n. 11; and Ridyard, *supra* n. 168.
[275] See Lind and Muysert, *supra* n. 37, 90, and the Charles River Report, *supra* n. 14. Note however a slightly more cautious approach by Hylton and Salinger, *supra* n. 4, pp. 470–471 and p. 525, and compare with Ahlborn, Evans and Padilla (2004), *supra* n. 2, 328–329 and 339–341.
[276] Ridyard, *supra* n. 168, 316.
[277] *US Microsoft III.*
[278] Lind and Muysert, *supra* n. 37, 90 and *US Microsoft III*, at 84–97.
[279] Lind and Muysert, *supra* n. 37, 90.
[280] Ibid.
[281] Ibid., 90.

Ridyard, on the other hand, argues that such a test based on an objective justification by the dominant company consisting of efficiency advantages gained by the tie outweighing any potential harm to competition, although less strict than the *per se* approach 'still places a burden of proof against normal commercial activity that has no justification in economics'.[282] Instead Ridyard suggests a test whereby one should 'identify the circumstances in which bundling can have adverse effects on competition, and then to work on operational rules that might help to distinguish these harmful cases from the generality of normal commercial behaviour'.[283] Unfortunately, Ridyard does not explain what these 'operation rules' should entail.

Kühn *et al.* offer a suggestion by setting up a three-step approach, based on economic assumptions and theories, which an effective rule on tying must satisfy: first, the rule must include a safe haven for those tying arrangements which do not cause anticompetitive effects. Second, it needs to create conditions which permit the assessment of anti-competitive effects to substantiate intervention. Third, the rule needs to allow for pro-competitive effects to be balanced against the anti-competitive effects.[284] Kühn *et al.* explain that the purpose of the safe haven rule is to eliminate the cases which pose no anti-competitive threat. According to their economic analysis, there are three criteria, all of which must be present before competition authorities should be concerned with the tying conduct:

(a)   market power in one of the tied product markets;
(b)   complementarity of the different products bundled together; and
(c)   asymmetry in product lines,

Kühn *et al.* argue that all foreclosure theories relating to tying depend upon some form of product line asymmetry between the tying company and its competitors.[285] Therefore 'where more products can be offered as part of the same bundle, there are greater opportunities for "Cournot effects" to arise, and this in turn greatly increases the profit incentive for bundling'.[286] Although this comment is not contentious, it must be emphasised that harm stems mainly from the market power element, rather than the asymmetry in the product lines. Therefore, foreclosure could also be likely without the product line asymmetry.

---

[282]   Ridyard, *supra* n. 168, 317.
[283]   Ibid.; as Ridyard himself mentions, this policy is consistent with an 'effect based' competition enforcement.
[284]   Kühn, Stillman and Caffarra, *supra* n. 11, 113.
[285]   Ibid., 114.
[286]   Ibid.

Kühn *et al.*'s 'safe haven' rule also includes those tying arrangements which have already been accepted by society and which cause no threat to competition such as selling shoes in pairs and cars. From a legal perspective, Kühn *et al.*'s argument makes sense. A 'safe haven' rule will hope to ensure needed legal certainty for companies. Moreover, such a rule corresponds well with current legal policy in other areas of competition law on both sides of the Atlantic. On the other hand, the safe haven rule will also permit the types of tying seen in the empirical evidence of the foreign electrical adapter market, which limit consumer choice.[287] The limiting of consumer choice may not on its own constitute an antitrust concern, but is closely associated with exploitative abuse, and the limitation of consumer choice can therefore work as an initial indicator of abuse.[288]

Kühn *et al.* state, 'the real debate is not about whether bundling can have anticompetitive effects. *It is about identifying the circumstances under which anticompetitive effects of bundling are likely to occur and what criteria should be used to come to the conclusion that an unbundling remedy is justified*'[289] (emphasis added). Although this line of reasoning compares well with the various arguments put forward as regards tying and its ability as a strategy to enable a company to expand market power or extend market power to adjacent markets, this is only so when certain circumstances are present. Hence, tying appears to cause limited harm when the tied product market is competitive: whereas when it is oligopolistic, tying can indeed aid a dominant company to deter entry into the tied product market, foreclose the market and exclude rivals.

Although reaching such a conclusion leads to the assumption that the scholars associated with the Chicago School were correct in arguing that tying should not be illegal *per se* under antitrust policy, it also emphasises the need for an accurate analysis of dominance and, not least, a proper definition of the relevant markets in question. If either is incorrect, the result can be that a company is punished for conduct which has no competitive harm, or alternatively a company can be left unpunished for tying conduct causing severe harm to competition on the tied product market, or maybe on both markets.

Ahlborn *et al.* identify a series of screens for whether competition authorities should assess tying arrangements.[290] Firstly, they highlight that anticompetitive effects are unlikely unless the company holds significant market

---

287   The CFI found this an antitrust concern in Case T–201/04 *EC Microsoft*, para. 945.
288   Monti, G., *EC Competition Law* (Cambridge University Press, Cambridge, 2007), 191.
289   Kühn, Stillman and Caffarra, *supra* n. 11, 88.
290   Ahlborn, Evans and Padilla (2004), *supra* n. 2, 330.

power in the tying market and there is imperfect competition in the tied market.[291] This conclusion is in line with the reasoning of other scholars above.[292] Ahlborn *et al.*, however, are more detailed than others, as they pinpoint seven conditions which must be met: (1) (significant) market power of the tying firm; (2) imperfect competition in the tied market; (3) commitment to the tie;[293] (4) competitors' inability to match the tie; (5) likelihood of competitors' exit; (6) barriers to entry; and (7) absence of buyer power.[294] If the tie does not satisfy these conditions, it is unnecessary to continue on to the second screen. That said, Ahlborn *et al.* suggest a certain amount of flexibility within these conditions and emphasise that it is really the first two points, significant market power and imperfect competition in the tied market, which weigh heavily in the conclusion that the tie may impose anti-competitive effects.[295]

In competition law, significant market power has become a well-known phenomenon and the courts and competition authorities have constructed viable elements to establish evidence of significant market power. However, in respect of imperfect competition, there is a risk that the competition authorities and courts will struggle – firstly to define the concept and secondly to prove it. The first question to ask is: what is imperfect competition? Sloman defines it as the collective name for monopolistic competition and oligopoly.[296] In other words, it is the market which operates neither under perfect competition conditions nor monopoly, two extremes which rarely, if ever, occur in the real world. Imperfect competition is the market condition in which the majority of industries work.[297] The requirement for imperfect competition will therefore be fulfilled more often than not, and hence it is not a particularly efficient measurement to apply in assessing whether the tying is likely to be anti-competitive or not.

A condition which is strikingly more likely to send warning signals as regards the tie would be the competitor's inability to match the tied product package, as this would indicate both high barriers to entry to one of the markets and risk of competitors exiting. There are likely to be only two reasons for this condition; firstly, it could be that the tie is creating a new inno-

---

[291]   Ibid.

[292]   See Kühn, Stillman and Caffarra, *supra* n. 11; and Carboja, De Meza and Seidman, *supra* n. 12.

[293]   As was shown in Whinston's model discussed above: Whinston (1990), *supra* n. 13, 843–848.

[294]   Ahlborn, Evans and Padilla (2004), *supra* n. 2, 331–333.

[295]   Ibid., 333.

[296]   Sloman, J., *Economics* (5th edn, Financial Times Prentice Hall (Pearson Education), Harlow, 2003), 150.

[297]   Ibid.

vative product; secondly, it could be that the tying company possesses monopoly or near-monopoly power in the tying market due to a product having become a *de facto* industry standard. Again in that case it would render it almost impossible for competitors to match the tying of the industry standard with another product. This is the case in both the *US Microsoft* cases and *EC Microsoft*, where Microsoft was found to have a market share of more than 90 per cent in the operating systems market.[298]

The second screen asks whether anti-competitive effects are plausible and here Ahlborn *et al.* argue for developing economic models to match the facts of individual cases. The models would then create a framework setting certain conditions which must be met for the tie to be eliminated from the process.[299] If this is not the case, the assessment moves to the third screen.[300] The second screen will eliminate tying arrangements as anti-competitive because the assumption of how these will create anti-competitive harm does not correspond with the facts of the situation.[301] An example of this can be seen in the analysis of the empirical evidence undertaken by Evans and Salinger described above.

Thirdly, there will be tying arrangements which reduce competition, and these should be assessed in a balancing test weighing the pro-competitive effects, such as costs reductions and quality improvement, against the anti-competitive effects, that is the reduction in competition.[302] This assessment also resembles the suggestions made above by Ridyard and Kühn *et al.*

This third screen requires the anti-competitive effects to be weighed against any efficiency benefits the tie may create.[303] This weighing needs to be done in the light of market conditions over time. In other words, dynamics and uncertainties must be taken into account as well.[304] Ahlborn *et al.* speculate that most tying arrangements, despite reaching the third screen, will still generate more efficiencies than anti-competitive effects. Their overall recommendation is a moderated legal *per se* rule (in particular for technological integration), and a *rule of reason* (mainly for contractual tying).[305] A modified *per se* approach is very similar to the approach currently applied in EC competition law, namely a special form of balancing test weighing up the pro- and anti-competitive effects. Although importantly the tying will be seen as legal

---

298   *US Microsoft III* at 51 and Case T–201/04 *EC Microsoft*, para. 31.
299   Ahlborn, Evans and Padilla (2004), *supra* n. 2, 334–336.
300   Ibid., 336.
301   Ibid., 330.
302   Ibid.
303   Ibid., 336.
304   Ibid.
305   Ibid., 339.

rather than the current illegal *per se* approach requiring more weight to be placed upon the balancing test of the pro- and anti-competitive effects.

Before concluding this examination of the economic analysis of tying, it is important to assess the material applied critically. The data consist of many joint articles by the same group of economic scholars. In particular, Evans, Salinger and Padilla seem to have been very active in this area, and thus their opinions are reflected throughout many of the articles. Their view is that the courts should look upon tying with more tolerant eyes. Their articles can be contrasted with those of Kühn *et al.* who hold a more traditional and cautious view. However, all recommend that tying should not be treated as always illegal *per se*.

The lessons to be learned from the economic experts are, firstly, that tying cannot be assessed by a simple rule such as the *per se* illegality rule under US antitrust law. This has also been recognised by the US courts and the EC Commission[306] and, as will be seen in Chapter 4, the actual *per se* approach adopted in the US to tying is much more flexible than strict illegality when tying conduct has been identified. In Europe, the test under Article 82 cannot be categorised as either an illegal *per se* test or a *rule of reason*; rather it is a structured balancing test, where certain elements, such as the risk of foreclosure and dominance, add more weight than others in comparison to economic efficiencies. The lessons drawn from the economic experts in respect of this test are that the scales would at least have to be re-adjusted to fit the economic findings.[307]

Secondly, it is clear that tying appears harmful only when the tying company possesses market power, and indeed a significant amount as when it has a monopoly or near-monopoly position. Hence, any form of competition rule which bans tying should do so only when market power exists. Thirdly, the reasons behind tying are plentiful, and thus a true assessment of such conduct and its consequences cannot be fully analysed without a review of the reasons behind the conduct. Naturally, such a review is difficult, as the company in question would attempt to justify the conduct by beneficial efficiencies rather than truthfully confess that the tie was introduced to exclude

---

[306]   See Microsoft: Statement by EU Commissioner Mario Monti (www.eurunion.org/news/press/2004/20040047.htm): '. . . [the] Commission has not ruled that tying is illegal per se, but rather developed a detailed analysis of the actual impact of Microsoft's behaviour, and of the efficiencies that Microsoft alleges . . . we did what the US Court of Appeals suggested be done: we used the rule of reason . . .'. In respect of the anti-competitive effects assessment the Commission did indeed apply a '*rule of reason*' approach; the rest of the Article 82 assessment, however, was a very traditional European *per se* illegality approach to tying (see the analysis in Chapter 3).

[307]   See Ridyard, *supra* n. 168, 319.

rivals. An opponent review is therefore essential to establish the (anti-competitive) effects that tying may have in the markets and on competitors. Only when the latter outweigh the former should the tie be condemned. Fourthly, a safe haven should be introduced to create legal certainty and, as Kühn *et al.* point out, to exclude the so-called socially acceptable ties. The simplest way to create such a safe haven is by allowing companies with *insignificant* market power to tie. This approach corresponds well with the findings of the other economists.[308] However, a safe haven for innovative and technologically complex products is also needed. Hence a safe haven rule needs also to be based on some form of product definition. Note, though, that Ridyard does not agree on this point: '[e]ffort spent on arguing over where one product ends and where a "bundle" begins is at best wasted, and more often than not is actually harmful to finding sensible consumer serving competition policy outcomes'.[309]

Finally, the above analysis, although highlighting some important aspects of tying and its anti-competitive effects in the markets, only reflects the economic perspective of tying and can thus merely suggest and not automatically determine how competition authorities should react to tying arrangements. At times, the courts have followed or applied a more economic approach to tie-ins. Nevertheless, as will be seen in the following chapters, this attitude is not the norm: courts have always sought and will most likely continue to seek their own independent view on matters, if only from a practical perspective. As Lessig notes, 'historically, antitrust law seems to . . . [condemn] the practice first, and only later coming to see that much of what it condemns is in fact competitively benign',[310] the reason being that it has first and foremost been the courts' obligation to apply the statutes to facts and ensure legal certainty.[311]

Overly complex cases with significant amounts of economic evidence are expensive, time-consuming and often difficult to derive legal certainty from. The courts may not be best suited to assess technical or economically difficult facts. Case law illustrates the courts' reluctance to delve into these issues, instead taking an easy opt-out.[312] Competition authorities are probably more

---

[308]   Such as Ahlborn, Evans and Padilla (2004), *supra* n. 2, and Evans and Salinger, *supra* n. 27.

[309]   Ridyard, *supra* n. 168, 319.

[310]   Brief of Prof. Lawrence Lessig as *Amicus Curiae*, in *United States v Microsoft Corp.* D.C. of Columbia, Civil Action No. 98-1233 (TPJ), 1 February 2000, 6–7.

[311]   Bork, *supra* n. 1, 72.

[312]   See *Telex Corp. v IBM Corp.*, *Response of Carolina, Inc. v Leasco Response, Inc.*, 537 F.2d 1307, 1330 (5th Cir. 1976); *ILC Peripherals Leasing. v IBM Corp. II and US Microsoft II* at 950 and footnote 13, in particular Judge Wald's dissenting opinion at 956–964.

likely to be influenced by economists, and may have a better feel for the more advanced technical and economic questions.[313] However, that still does not necessarily make them well suited to assessing, for instance, whether a product is truly integrated or not.

Even if a more refined test for tying is adopted it needs to also take the following aspects into consideration. The courts will be faced with complex questions regarding the standard of proof placed on both the plaintiff and the defendant. This issue becomes even more crucial as most plaintiffs in private actions are rival companies and not harmed consumers. As Ridyard argues '. . . whilst adverse outcomes for consumers can result from the exclusion of competitors, the incentive for rivals to complain bears no reliable relationship to the risk that the [tying] will harm competition'.[314] The economists above offer in principle a balancing test for tying arrangements. However, such a test requires detailed and vigorous assessment of both pro- and anti-competitive effects, an assessment which is difficult to carry out with real world conditions, as well as *ex ante*.[315] Therefore, a more structured analysis must be sought, one which can screen out pro-competitive tying arrangements early on and allow for a more thorough but effective assessment of those tying arrangements which pose a threat.

Article 82 contains such a structured balancing test, firstly, only targeting conduct by dominant companies and, secondly, containing an assessment of anti-competitive effects and allowing for objective justifications to be taken into consideration. However, some elements triumph over others, just as anti-competitive effects outweigh objective justifications due to an imbalance in the standard of proof between the two. US antitrust case law seems also to have introduced a similarly structured balancing test for tying arrangements. Both approaches will be assessed to identify a more dynamic approach to tying arrangements which takes into consideration the findings in this chapter, as well as ensuring all-important legal certainty.

---

[313]   For instance the recent modernisation of Article 81 and the quest for modernising Article 82 by the European Commission; another is the change in the US Department of Justice's policy towards licensing of IP rights.

[314]   Ridyard, *supra* n. 168, 319.

[315]   Melamed, A.D., 'Exclusionary conduct under the antitrust laws: balancing, sacrifice, and refusal to deal' (2005) 20 *Berkeley Technical Law Journal* 1247, p. 1254.

3. Tying Arrangements under Article 82 EC

1. INTRODUCTION

Two important elements create the framework of any Article 82 analysis: dominance and abuse. However Article 82 falls short of providing a definition of the two concepts. Therefore, case law has played a crucial part in shaping the role that Article 82 takes in EC competition law regulation. Without tracing the whole history of Article 82, its roots, and case law development, it suffices to say that Article 82 has been heavily influenced, if not based, on ordoliberal thinking and therefore began as a means to control market power, to create fair competition and ensure the freedom of those engaged in business.[1] Out of this backdrop have evolved certain central concepts such as the special responsibility placed upon dominant companies and performance-based competition, which have led to certain conduct being treated as more or less illegal *per se*.[2] This is regardless of the actual efficiencies and benefits the particular conduct generates for consumer welfare, because the current form-based approach merely identifies and condemns specific conduct as abusive and presumes the (potential) effects upon competitors.[3] The result is an increased risk of false positive rulings, where conduct appears anti-competitive from a pure formalistic view, but in reality may not be as anti-competitive and has significant efficiency and welfare benefits,[4] tying being an excellent example. Case law has

---

[1]   Gerber, David, *Law and Competition in the Twentieth Century Europe, Protecting Prometheus* (Oxford University Press, Oxford, 2001), p. 241 and Rousseva Ekaterina, 'Modernising by eradicating how the Commission's New Approach to Article 81 EC dispenses with the need to apply Article 82 to vertical restraints' (2005) 42 *CMLRev.* 587, p. 589.

[2]   Ibid., 592–595 and Ridyard, Derek 'Exclusionary pricing and price discrimination abuses under Article 82 – an economic analysis' (2002) 23 *European Competition Law Review* 286, at 290–291.

[3]   Padilla, Jorge, 'The Reform of Article 82: What we agree, what we are still discussing and what will have to be discussed', Speech for LECG, Paris, 3 July 2007 (www.lecgcp.com/resources/documents/Library_300%20-%20The%20reform%20of%20article%2082-2.pdf?pubtitle=LECG).

[4]   Ibid.

demonstrated that tying has been treated as *almost* illegal *per se*, whereas economic theory suggests that only under specific circumstances is tying actually harmful for consumer welfare.

In 2005 the Commission launched a review of Article 82.[5] This culminated in December 2008 with the publication of a set of guidance notes on the Commission's enforcement priorities in applying Article 82 to exclusionary abuses, the *Article 82 Guidance*.[6] The review should be seen in conjunction with the modernisation that the other areas of EC competition law went through prior to 2005, and the possibility of streamlining it with these. The starting point was therefore a more economic approach: one which could both accommodate recent economic thinking by being more realistic or true to the reality of the market place, and take into greater consideration the pro- and anti-competitive effects of the conduct to determine whether it violates Article 82,[7] but also contain practical and workable rules.[8] The main concern therefore lies with the interpretation of abuse and how to distinguish this from competition on the merits.[9] Yet the finding of dominance is in effect what triggers the use of Article 82, and thus has also played a significant role in Article 82 case law.

Both Articles 81 and 82 prohibit tying.[10] This is evident from the almost

---

[5]   Commission, 'Article 82 Review' (http://ec.europa.eu/competition/antitrust/art82/index.html).

[6]   'Guidance on the Commission's Enforcement Priorities in Applying Article 82 EC Treaty to Abusive Exclusionary Conduct by Dominant Undertakings, Brussels, 3 December 2008 (http://ec.europa.eu/competition/antitrust/art82/guidance.pdf) (the *Article 82 Guidance*), p. 4.

[7]   See Economic Advisory Group on Competition (EAGCP), 'An Economic Approach to Article 82', July 2005 (http://ec.europa.eu/comm/competition/publications/studies/eagcp_july_21_05.pdf); Vickers, John, 'Abuse of Market Power', chapter 11 in Buccirossi, Paolo (ed.), *Handbook of Antitrust Economics* (MIT Press, Cambridge, Mass., 2008), and British Institute of International and Comparative Law, Competition Law Forum, 'The Reform of Article 82: Comments on the DG-Competition Discussion Paper on the Application of Article 82 to Exclusionary Abuses', London 31 March 2006, (http://ec.europa.eu/comm/competition/antitrust/others/054.pdf).

[8]   Commission, *supra* n. 5.

[9]   See O'Donoghue, Robert and Padilla, A. Jorge, *The Law and Economics of Article 82 EC* (Hart Publishing, Oxford and Portland, Oregon, 2006), p. 176.

[10]   Some commentators have argued that it would be more convenient if the EC competition law changed to follow the US approach and made Article 81 deal exclusively with tying: Monti, Giorgio, 'DG Competition Discussion Paper on the application of Article 82 of the Treaty to Exclusionary Abuses (December 2005), Comments from: Giorgio Monti, Lecturer in Law, Law Department, London School of Economics' (http://ec.europa.eu/comm/competition/antitrust/art82/065.pdf), p. 5. and Rousseva, (2005) *supra* n. 1.

identical lists of prohibited conduct/abuses in both Articles. Included in these lists is tying, where making 'the conclusion of contracts subject to acceptance by the other parties of supplementary obligations which, by their nature or according to commercial usage, have no connection with the subject of such contracts'[11] is prohibited. Not only does this suggests that tying is seen as conduct which can cause severe harm to competition, but also that it is a floating concept in the sense that tying can be used by companies as part of an agreement, as unilateral conduct, or even as a strategy when merging.[12] Tying is thus caught under several provisions within EC competition law. The definition describes a contractual tying arrangement, and thus on the surface the scope of abuse is rather narrow in comparison to that of the other examples given under Article 82.[13] However, the Court of First Instance (CFI) and the European Court of Justice (ECJ) (referred to collectively as the EC Courts) and the European Commission (the Commission) have not limited themselves to this definition in case law, but have applied Article 82 much more widely, and there appears to be no distinction in their assessment of the different types of tying arrangements.[14]

---

[11]   Articles 81(e) and 82(d) of the EC Treaty.

[12]   See Case no. COMP/M 2220 *General Electric/Honeywell*, Commission Decision, 03 July 2001 (http://ec.europa.eu/comm/competition/mergers/cases/decisions/m2220_en.pdf).

[13]   O'Donoghue and Padilla, *supra* n. 9, p. 206. Another category of bundling is that of package licences, where it is patent licences rather than products that are bundled together. Package licensing is very similar to tying as it also raises the risk of foreclosure, requires the identification of the same elements as tying such as coercion, market power and the identification of at least two separate products, and they are therefore treated very similarly to tying of ordinary products under both the IP misuse doctrine and competition law: Hovenkamp, Herbert, Janis, Mark D., and Lemley, Mark A., *IP and Antitrust, An Analysis of Antitrust Principles Applied to Intellectual Property Law* (Aspen Law & Business, New York, 2004), Chapter 22.

[14]   Case T–201/04 *Microsoft Corp. v Commission*, [2007] ECR II-3601, paras 860–861 and Case COMP/ C–3/37.792 *Microsoft*, [2005] 4 CMLR 965, paras 794 and 801, The *Article 82 Guidance*, *supra* n. 6, at 18, distinguishes between contractual tying and technological integration, but merely note that the foreclosure effects are greater for technological integration; compare with DG Competition, Commission, *DG Competition discussion paper on the application of Article 82 of the Treaty to exclusionary abuses* (Brussels, December 2005), (http://ec.europa.eu/comm/competition/antitrust/art82/discpaper2005.pdf), p. 55. The paper is a Staff Discussion paper published by the DG Competition and not actually an official Commission Notice. For simplicity however, it will be referred to as the Commission *Discussion Paper*. See also Faull, J. and Nikpay, A. (eds), *The EC Law of Competition* (Oxford University Press, Oxford, 2007), p. 370; Langer, J., 'Bundling, A Four-step Test to Assess the Exclusionary Effects of Bundling under Article 82 EC' in Amato, G. and Ehlermann, C-D., *EC Competition Law, A Critical Assessment* (Hart Publishing, Oxford, 2007),

This means in practice that Articles 81 and 82 may apply to the same tying conduct or agreement, but on their own terms.[15]

Notably little tying case law exists under Articles 81 and 82.[16] The case law found can be divided into three categories: (1) disguised or indirect forms of tying agreements, which are mainly grants of rebates to overall purchases;[17] 2) tying clauses in licensing agreements;[18] and (3) tying as unilateral conduct by a dominant undertaking.[19] The first category has shown a mixture of both Article 81 and Article 82 cases, whereas the second and third categories apply only to Article 81 and 82 cases respectively. It is also the latter two categories that contain the most cases of tie-ins with IP right-protected products.[20] Although there are only a handful of cases in the third category, it has aroused the most controversy and debate, and interestingly most of these involved some form of IP rights.

The recent modernisation of Article 81 has meant that the approach applied to vertical restraints such as tying under Article 81 is more economic and focuses on consumer welfare, compared to the still form-based and rigid

---

p. 319; O'Donoghue and Padilla, *supra* n. 9, p. 491; see also Korah, Valentine, *Intellectual Property Rights and the EC Competition Rules* (Hart Publishing, Oxford and Portland, Oregon 2006), pp. 157–159 for a slightly different view.

[15]   Anderman, Steve, *EC Competition Law and Intellectual Property Rights, the Regulation of Innovation* (Oxford University Press, Oxford, 1998), p. 147, footnote 1.

[16]   Waelbroek, Denis, 'The Compatibility of Tying Agreements with Antitrust Rules: A Comparative Study of American and European Rules' [1987] *Yearbook of European Law* 39, p. 48 and Ahlborn, Christian, Evans, David S., and Padilla, A. Jorge, 'The Antitrust Economics of Tying: A Farewell to *Per Se* Illegality', *Antitrust Bulletin*, Spring–Summer 2004, 287, p. 306.

[17]   Case 85/76 *Hoffmann–La Roche v Commission* [1979] ECR 547, para. 111 and Case 322/81 *Michelin v Commission* [1983] ECR 3520 (*Michelin I*), para. 92 et seq.

[18]   *Vaessen-Moris*, Commission Decision, 10 January 1979 [1979] OJ L19/32, [1979] 1 CMLR 511 and Case 193/83 *Windsurfing International Inc. v Commission* [1986] ECR 611, [1986] 3 CMLR 489; however, most of these are now covered under Commission Regulation (EC) No 772/2004 of 27 April 2004 on the application of Article 81(3) of the Treaty categories of technology transfer agreements [2004] OJ L123/11 (the TTBER) under Article 81.

[19]   Case C–53/92P *Hilti AG v. Commission* [1994] ECR I–666; [1994] 1 CMLR 590, Case T–30/89 *Hilti AG v Commission* [1991] ECR II–439, [1992] 4 CMLR 16; Case IV/30.787 *Eurofix-Bauco/Hilti* [1988] OJ L65/19 (*Hilti* Commission Decision) and Case C–333/94 *Tetra Pak International SA v Commission* (*Tetra Pak II*) [1996] ECR I–5951, [1997] 4 CMLR 662; Case T–83/91 *Tetra Pak Rausing SA v Commission* (*Tetra Pak II*) [1990] ECR II–309, [1991] 4 CMLR 334; and *Elopak Italia/ Tetra Pak* [1991] OJ L72/1, [1992] 4 CMLR 551 (*Tetra Pak II* Commission Decision); see also Waelbroeck, *supra* n. 16, p. 48.

[20]   For licensing agreements see Case 193/83 *Windsurfing International* and Commission Decision *Vaassen-Moris*. For unilateral conduct: Cases C–53/92P *Hilti AG* and C–333/94 *Tetra Pak II*.

approach used under Article 82 creating a precarious inconsistency in the interpretation of the same conduct.[21] For instance tying has not been labelled a hardcore restriction in the Technology Transfer Block Exemption Regulation (the TTBER),[22] signifying that the Commission, when it comes to licensing agreements, has found it less harmful, especially without the presence of market power. Moreover, a more flexible and more an effect-based approach has been applied to other forms of conduct under Article 82, such as refusal to supply.[23] There is therefore good reason to assess the approach applied to tying under Article 82 and ask whether a better, more realistic and flexible approach can be adopted – one that takes into account the benefits of tying and aligns itself with recent modernisation?

2. ASSESSING TYING UNDER ARTICLE 82

Tying can be a leveraging device. This means that there are at least two markets involved, and in case law it has been presumed that the tying conduct will affect these markets in a *per se* anti-competitive manner, although economists have proven that tying is not always anti-competitive, but can induce pro-competitive efficiencies as well.[24] These underlying issues have all participated in shaping the test applied for tying in EC case law.

However, in comparison to US antitrust law, the approach to tying is also shaped on the basis of Article 82 and not just case law. This means that case law adheres to a rigid approach requiring firstly the definition of the relevant market, then the establishment of dominance, before the assessment of the alleged abusive conduct is undertaken. However, for tying, the abuse analysis requires a return visit to the relevant market definition and the establishment of dominance, because a tying arrangement can infringe Article 82 only if there are two products present and dominance in the tying product market. This section focuses only on the abuse assessment of tying and will follow the five steps below:

---

[21]   Commission Notice: Guidelines on the Application of Article 81(3) of the EC Treaty [2004] OJ C101/97; IP/00/520, 'Commission finalises new competition rules for distribution', Brussels, 24 May 2000; and Rousseva *supra* n. 1 (2005), p. 589.

[22]   The TTBER, *supra* n. 18, Article 4.

[23]   Case C–7/97 *Oscar Bronner GmbH & Co KG v Mediaprint Zeitungs- und Zeitschriftenverlag GmbH & Co KG* [1998] ECR I–7791, [1999] 4 CMLR 112 and C–418/01 *IMS Health GmbH & Co. OHG v NDC Health GmbH & Co. KG* [2004] ECR I-5039, [2004] 4 CMLR 1543.

[24]   See Chapter 2 and O'Donoghue and Padilla, *supra* n. 9, pp. 206–207.

(1)   defining the relevant market and the existence of a tied product separate from the tying product;
(2)   the undertaking concerned is dominant (normally in the tying product market, but can also be found dominant in the tied product market[25]);
(3)   coercion, that is forcing customers to purchase the bundle;[26]
(4)   a foreclosure effect on competition due to the tie-in;
(5)   no objective justification for the coercion.[27]

2.1. Defining the Relevant Market – the Need for Two Separate Products

The market definition assessment applied by the EC Courts and the Commission is of critical importance to whether or not a company is found guilty of tying, as one of the main criteria for abuse by tying is that there are in fact two separate products. This means that there is a powerful link between the analysis of the relevant market, which the Commission and the EC Courts embark upon in all Article 82 cases, and the (potential) finding of a separate product tied to another product, which is specific to tying abuse cases. A tying case can thus quickly be lost or won based on the determination of the relevant product market(s). For instance, through the entire objective justification assessment in *EC Microsoft*, the Commission rejected most of Microsoft's defences because it had previously found two separate products and not, as Microsoft had argued, an integrated unit.[28]

The relevant market definition works as a framework for the establishment of whether the company holds a dominant position, as a dominant position only exists in relation to a market.[29] The market analysis consists of two

---

[25]   *IBM Undertaking* [1984] OJ L118/24.
[26]   The *Article 82 Guidance*, *supra* n. 6, at 18, does not mention coercion, the tying merely 'is likely to lead to anticompetitive foreclosing'.
[27]   The above elements are based on the different versions/approaches given by the Commission and scholars, but are an expression of the author's own interpretation of the test applicable for tying under Article 82. See Case T–201/04 *EC Microsoft*, paras 839–871; *EC Microsoft* Commission Decision, para. 794, Commission *Discussion Paper*, *supra* n. 14, p. 55; *Article 82 Guidance*, *supra* n. 6, p. 18; and Dolmans, M. and Graf, T., 'Analysis of Tying Under Article 82 EC: The European Commission's Microsoft Decision in Perspective' (2004) 27 *World Competition* 225, p. 226, Diaz, F.E.G. and Garcia, A.L., 'Tying and bundling under EU competition law: future prospects' (2007) 3 *Competition Law International* 13, pp. 13–14; Ahlborn, Evans and Padilla, (2004), *supra* n. 16, pp. 310–311.
[28]   *EC Microsoft* Commission Decision, *supra* n. 27, paras 956–970.
[29]   Case 27/76 *United Brands v Commission* [1978] ECR 207, para. 11; Case T–62/98 *Volkswagen AG v Commission* [2000] ECR II–2707, para. 230; and Commission *Discussion Paper*, *supra* n. 14, p. 9. See also Commission Notice on the

elements: the finding of the relevant product market and the finding of the relevant geographic market.[30] The former element is divided into two further parts: demand substitutability and supply substitutability.[31] The definition of the relevant market revolves around identifying products which are substitutes. In EC case law, the assessment has been based on two tests: a subjective test, which looks at the products' characteristics, intended use and price,[32] and a more objective economic test, the SSNIP test,[33] which measures consumers' and rival producers' reaction to a price increase of a maximum of 10 per cent of one product.[34] Article 82 cases may also involve markets where there is no company holding a dominant position, but a company from another market is leveraging its market power into the market.[35] In such instances, it is necessary to identify both markets. This is of course the case with tying.

The second part of the relevant market analysis requires the identification of the relevant geographic market. The Commission defines the geographic market as an area in 'which the conditions of competition are sufficiently homogeneous and which can be distinguished from neighbouring areas because the conditions of competition are appreciably different in those areas'.[36] In assessing this, the Commission has looked at cost of transport, the nature of the product and legal regulations.[37]

2.1.1. Defining the relevant market in tying cases

In *Hilti*, Hilti manufactured nail guns, cartridge strips for which it held patents for (and some parts of the cartridge strips were covered by copyright in the UK).[38] Together with nails, which it also produced but held no IP right over, it sold them in a package; the powder-actuated fastening system (PAFS). In

---

definition of the relevant market for the purposes of Community competition law [1997] OJ C372/5, [1998] 4 CMLR 177, paras 7 and 8.

[30]   Ibid., when necessary a third dimension temporal market is applied as well. This is not mentioned in the Commission Notice, but applied for instance in Case 27/76 *United Brands*.

[31]   Commission Notice on the definition of the relevant market, *supra* n. 29, paras 13–23.

[32]   Case 27/76 *United Brands*, paras 31 and 34.

[33]   SSNIP (Small but Significant Non-transitory Increase in Price) also called price-cross elasticity of demand for a product: Commission Notice on the definition of the relevant market, *supra* n. 29, para. 17.

[34]   The price measured from is assumed to be the competitive price.

[35]   The Commission *Discussion Paper*, *supra* n. 14, p. 8.

[36]   Commission's Notice on the definition of the relevant market, *supra* n. 29, para. 8.

[37]   See for instance Commission Decision *British Midland/Aer Lingus* [1992] OJ L96/34, [1993] 4 CMLR 596; Case 322/81 *Michelin I*, and Commission Decision *British Telecommunications* [1982] OJ L360/36, [1983] 1CMLR 457.

[38]   Case T–30/89 *Hilti*, para 93.

addition to Hilti, there were some small independent manufacturers producing Hilti-compatible nails. Hilti made use of several practices to ensure that customers buying its cartridges would also purchase its nails rather than the independent manufacturers' nails. One of these was forcing the customers also to purchase a set number of nails when purchasing its patented cartridge strips. Upon investigation, the Commission found firstly that there were three separate markets: one for the nail guns, one for the cartridge strips, and one for the nails, and not, as Hilti had argued, one single market for the PAFS.[39] Three arguments swayed the Commission. First, Hilti held a patent on the cartridge strips; because the cartridge strips required to be specifically adapted to fit the nail gun, the patent legally allowed Hilti to exclude competition, thereby giving Hilti market power.[40] Second, Hilti's position was further empowered by its technological advantages in its products, R&D and a well-organised distribution system in the nail gun market.[41] Third, there were independent nail manufacturers. Therefore, a separate demand, or rather different demand conditions, existed for nails in comparison to the PAFS.[42] Importantly, the Commission noted that the nail guns and consumables (nails and cartridge strips) were not purchased together.[43] The nails therefore formed part of an aftermarket to the nail gun rather than a true component of a PAFS as a product. Moreover, there was an equal demand for both cartridges and nails, as each cartridge could contain just one nail.[44] According to Bishop and Walker, extracting two monopoly profits in such markets is not plausible.[45] The Commission nevertheless argued that cartridges and nails were in separate markets because different technologies and firms were involved in their production, and although an equal complement of nails and cartridges were needed, they were not necessarily purchased in equal quantities.[46]

The CFI supported the Commission's findings and the narrow definition of the relevant product market. However, Advocate General Jacobs, the Advocate General for the case before the ECJ, questioned the established market definitions by noting that if it was found that there was a significant degree of substitutability between PAFS and other similar systems, then Hilti would not

---

[39]   Ibid., para 68.

[40]   *Hilti* Commission Decision, paras 55 and 66.

[41]   Ibid., para 67.

[42]   Ibid., paras 57 and 66–67.

[43]   Ibid., para 57.

[44]   Nalebuff, Barry, 'Bundling, Tying, and Portfolio Effects', DTI Economics Paper No 1, Part 1 – Conceptual Issues, February 2003, pp. 17–18.

[45]   Bishop, Simon and Walker, Mike, *The Economics of EC Competition Law, Concepts, Application and Measuments* (Sweet and Maxwell, London, 2002), para. 6 p. 56.

[46]   *Hilti* Commission Decision, para 55.

be dominant.[47] AG Jacobs concluded though by accepting the CFI's finding that there was low substitutability between PAFS and other fastening systems.[48] The ECJ decided not to look further into the relevant market definition, since the CFI had already accepted the Commission's findings and because it was not its responsibility as an appellate court.[49] The case makes it clear that the two determining factors for the relevant market definition and therefore also the establishment of dominance were the presence of independent manufactures in the nail market and the IP right held by Hilti.[50]

A similar approach was taken by the Commission and the EC Courts in *Tetra Pak II* where they also adopted a narrow market definition based upon the scope of the IP right rather than the product definition applied by Tetra Pak. Tetra Pak was licensing carton machines for aseptic and non-aseptic packaging, but forced its licensee to purchase the cartons from Tetra Pak and to obtain all maintenance, servicing and spare parts from Tetra Pak. Tetra Pak held patents for the basic technology of its machines, cartons and processes, and also on subsequent modifications, and certain techniques, such as the method of folding the carton.[51] Tetra Pak saw the machines and cartons it delivered and licensed as an integrated packaging system for packaging liquid food.[52] The Commission, however, identified four separate markets: the markets for aseptic cartons and for non-aseptic cartons, and the markets for aseptic machines and for non-aseptic machines.[53] The CFI concurred and noted that unless there is an infringement of IP rights, independent manufacturers are 'quite free' to manufacture consumables intended for use in other manufacturers' products.[54]

One consequence of this approach in *Hilti* and *Tetra Pak II* is that product systems such as Tetra Pak's packaging system and Hilti's PAFS can be seen as products only if all components are protected under the same IP right.[55] The tying of patented and unpatented products, which in effect goes beyond the material scope of the exclusive right a patent grants, is an infringement of

47   Case C–53/92P *Hilti AG v Commission*, AG Jacobs opinion [1994] 4 CMLR 614, p. 624.

48   Ibid., pp. 625–626 and Case T–30/89 *Hilti*, para 71.

49   Case C–53/92P *Hilti*, paras 11–16.

50   Case T–30/89 *Hilti*, para 67 and Govaere, I., *The Use and Abuse of Intellectual Property Rights in EC Law* (Sweet and Maxwell, London, 1996), para. 5 p. 36.

51   Case T–83/91 *Tetra Pak II*, para. 10.

52   Ibid., para. 34.

53   Ibid.

54   Ibid., para 83–84 and see also Case T–30/89 *Hilti*, para 68.

55   Anderman, *supra* n. 15, p. 157.

Article 82.[56] This assumption is very similar to that made in some US patent misuse cases,[57] but ignores the issue raised in other US cases where a distinction was made between staple and non-staple products. If innovation and product development are the core of an industry such policy can be very damaging to the progress of the industry and may overemphasise the market power conferred by the IP right.

Interestingly, both *Hilti* and *Tetra Pak II* have been classified as contractual tying cases, meaning that the tying arrangement involved an actual restriction on the customer's choice of where to obtain the tied product.[58] These arrangements can be distinguished from technological integration, where products are sold in an integrated unit.[59] Yet the Commission seems to have ignored this distinction and treated all tying arrangements the same, that is finding them all to infringe Article 82.

In *IBM Undertaking*, the Commission started proceedings against IBM after having received complaints from independent software companies alleging that IBM had abused its dominant position by tying its mainframe memory and usable software together. The Commission identified three markets: the core market for mainframe computer systems within which IBM's System 370 Central Processing Unit (CPU) was a product, and two submarkets: the market for main memory attached to the System 370 CPU and the market for software usable in the System 370 CPU. The reasoning behind the narrow market definition was that once customers had chosen the System 370 CPU, they were locked into it.[60] Consequently, IBM was found to be dominant in both submarkets. However, informal discussions between IBM and the Commission led to a settlement before the Commission reached a formal decision:[61] IBM agreed to offer its mainframe computer CPUs without memory devices. The case is yet another example of the Commission applying the market definition as a tool to establish dominance. The Commission did not alter its approach even though the case dealt with technologically integrated products.

In *EC Microsoft*, Microsoft had allegedly bundled its Windows Media Player (hereafter WMP) into the Windows operating system,[62] thereby allowing it to use its near-monopoly position in the market for desktop operating systems to harm its competitors in the market for audiovisual players, such as

---

[56]   Govaere, *supra* n. 50, para. 5.40.
[57]   See *Mercoid Corp. v Minneapolis-Honeywell Regulator Co.* 320 U.S. 680 (1944), at 684.
[58]   O'Donoghue and Padilla, *supra* n. 9, p. 492.
[59]   Cases such as *IBM Undertaking*, and *EC Microsoft* falls under this category.
[60]   Anderman, *supra* n. 15, p. 175.
[61]   O'Donoghue and Padilla, *supra* n. 9, pp. 495–496.
[62]   Case T–201/04 *EC Microsoft*, paras 860–861.

RealNetworks RealPlayer and Apple Computer Quicktime,[63] to the extent that there was a risk that if Microsoft continued to tie its media player to its operating system the other media player providers would be forced out of the market.[64]

The CFI did not assess the relevant market, but merely accepted and agreed with the Commission's findings.[65] The Commission had found two product markets: the operating systems market and the media player market,[66] because there was a separate consumer demand for media players from operating systems, by virtue of the existence of stand-alone media players which could be installed on PCs and vendors specialising in media players.[67] The Commission therefore disagreed with Microsoft's argument that the WMP was essentially an integrated component of Windows, and thus not a separate product.[68] At the same time, it rejected Microsoft's argument that an operating system without a media player was not an attractive product for consumers.[69] The Commission argued that because there were barriers to entry such as network effects and IP rights in the media player market, the value of a media player would be minuscule without corresponding digital content making it extremely difficult for new entrants to the markets.[70]

Hilti, Tetra Pak II, IBM Undertaking, and *EC Microsoft* demonstrate a tendency by the Commission and the EC Courts to identify the markets narrowly, which naturally leads to a greater likelihood of finding dominance in one of the markets. This affects tying arrangements cases in two respects: firstly, the requirement of dominance in the tying product market, which is more likely to be established with a narrow market definition, and secondly, the two products requirement, where narrow market definitions will assist in determining the presence of two products, rather than two components of a product.

2.1.2. The need for two separate products

The test applied to ascertain whether there are two separate products has proved controversial.[71] It is based around the question whether there is a

---

63 Lawsky, David, 'EU Set to Rule Against Microsoft', Reuters, 23 March 2004 (www.reuters.com).
64 *EC Microsoft* Commission Decision, para. 842.
65 Case T–201/04 *EC Microsoft*, paras 23–29.
66 *EC Microsoft* Commission Decision, para. 825.
67 Ibid., paras 405 and 804.
68 Ibid., para. 800.
69 Ibid, paras 404–405.
70 Ibid., paras 418–423.
71 Diaz and Garcia, *supra* n. 27, p. 13.

consumer demand for the tied product separate from the tying product.[72] The test revolves round identifying firstly the product's nature and physical features, and its commercial usage,[73] and, secondly, a consumer demand for the tied product, identified by whether there are any independent manufacturers in the tied product market.[74] This test therefore confirms the conclusion reached in Chapter 2 that a safe haven for certain tying arrangement could be introduced when products consist of integrated components rather than tied products, leaving these products completely outside the scope of Article 82.[75]

The finding of independent manufacturers in the alleged tied product market has been the benchmark of the Commission and the EC Courts.[76] In *Hilti*, the CFI held that the presence of independent producers in itself was 'sound evidence' of a separate 'Hilti-compatible nails' market.[77]

Two other tests can indicate the stand-alone status of a product: 1) an assessment of the company's own commercial conduct that is whether it sells, advertises or promotes the product separately, and 2) attempting to match the bundle offered by acquiring the components as separate products from other manufacturers and then assessing whether these are interchangeable with the bundle.[78] The former was applied in *EC Microsoft* by the Commission together with the finding that there were independent manufacturers and seemed to have played an important role in the finding of a separate market for media players.[79] The CFI also applied the latter comparative assessment, as will be discussed further below.[80] Although Microsoft sold its Windows operating system with its WMP, it also promoted it, offered upgrades separately from the operating system, and used different licensing conditions for the WMP from those it used for Windows.[81] Finally, Microsoft offered the WMP to other operating systems such as Apple.[82] These business strategies by Microsoft indicate that Microsoft itself knew and accepted that WMP was not

---

[72]   Dolmans and Graf, *supra* n. 27, p. 227, see also Case T–201/04 *EC Microsoft*, para. 917.

[73]   Ibid., para. 925 and Diaz and Garcia, *supra* n. 27, p. 13.

[74]   Case C–333/94 P *Tetra Pak II*, para. 36, T–30/89 *Hilti*, para. 67, T-201/04 *EC Microsoft*, para. 927, *Article 82 Guidance*, *supra* n. 6, p. 18, and Diaz and Garcia, *supra* n. 27, p. 13.

[75]   Chapter 2, Section 7.

[76]   Cases C–333/94 P *Tetra Pak II*, para. 36, T–30/89 *Hilti*, para. 67, T–201/04 *EC Microsoft*, para. 927, and *Article 82 Guidance*, *supra* n. 6, p. 18.

[77]   Case T–30/89 *Hilti*, para 67.

[78]   Dolmans and Graf, *supra* n. 27, pp. 227 and 239.

[79]   *EC Microsoft* Commission Decision, paras 805, 810, 813, and 821–824.

[80]   Case T–201/04 *EC Microsoft*, paras 921–922 and 943.

[81]   *EC Microsoft* Commission Decision, paras 805 and 813.

[82]   Dolmans and Graf, *supra* n. 27, p. 239.

a fully integrated component of Windows.[83] Moreover, originally WMP was distributed separately from Windows until Microsoft tied the products together.[84]

However, perceptions of products change over time, and thus in today's world a consumer would most likely expect that an installed operating system can play both CDs and DVDs without requiring further downloads or installation of further programmes.[85] In *EC Microsoft*, the Commission recognised the possibility of change in consumer perception of a product, and even criticised the consumer demand test applied in US case law, because it did not take into account benefits gained from product integration.[86] The Commission nevertheless found that '. . . there is *non-insignificant* consumer demand for alternative [media] players some four years after Microsoft started tying its [WMP] with Windows'[87](emphasis added).

Even though the Commission acknowledged that consumer perception changes, it did not acknowledge that this may be a process which does not occur overnight; hence there will be a demand for both options for a number of years, until the market fully adjusts. It was thus aware of the specific market conditions surrounding the case, but dismissed these as unimportant for the two separate products assessment.

A similar example is that of mobile phones with cameras. Although mobile phones with and without cameras are still competing against each other in the market, there is no doubt that the phones without cameras are being phased out as consumers naturally prefer phones offering the latest technologies. However, there are no demands upon the mobile phone manufacturers to permit other manufacturers to supply the camera function on the phone. This is partly due to the acceptance of integration of the two products and partly due to the market being competitive.

Undoubtedly, consumers' perception of products change. Demand for two separate products can disappear if an integrated version becomes popular, but equally where a component of a product generates particular benefits, a separate consumer demand for that component can also develop.[88] In which direction demand moves should be left to the consumers, and not be dictated by the conduct of a dominant company (or a competition authority for that matter).[89]

---

[83] Ibid.

[84] Ibid.

[85] See *EC Microsoft* Commission Decision, paras 404 and 809, and Chapter 2, Section 6, 'Empirical Evidence'.

[86] *EC Microsoft* Commission Decision, para. 808.

[87] Ibid., para. 808.

[88] Dolmans and Graf, *supra* n. 27, p. 228.

[89] Ibid.

In its *Discussion Paper* the Commission stated that it would consider technological integration as a possible defence for tying and it would do so with a futuristic perspective,[90] but would require that the independent demand for the tied product had ceased.[91] In comparison, the *Article 82 Guidance* states that as long as 'a substantial number of customers' are willing to purchase the tying product alone, the two products are distinct.[92] This definition of distinct products appears to be more lenient than the one suggested in the *Discussion Paper* and the CFI's *EC Microsoft* ruling.[93] Yet, it is in line with the Commission's approach towards technology transfers: 'it is a condition that the products and technologies involved [in the transfer] are distinct in the sense that *there is a distinct demand for each of the products and technologies forming part of the tie or the bundle*'[94] (emphasis added). A similar requirement is made in *Jefferson Parish* and *Eastman Kodak*, in the US of the consumer demand being sufficient for it to be economically viable to offer the products separately.[95] It is unclear how much can be read into the definition given in the *Article 82 Guidance*, because the definition bases itself upon the very paragraphs of the CFI's ruling which, in the opinion of the author, offer a much stricter reading of independent demand; namely only 'in the absence of independent demand' can there be no tie and thus no abuse.[96] Moreover, the *Article 82 Guidance* also makes clear that although technological integration would still be considered as a defence for tying, the Commission believes that the anticompetitive effects generated from such a permanent tie can be greater than for contractual tying, indicating that technological integration should be treated more strictly than contractual tying.[97]

The CFI concurred with the Commission's findings in *EC Microsoft*[98] and, judging from the Commission's statements above, this stringent test for product integration is likely to be upheld, setting a very high standard for accepting change in the market.[99] The CFI stated that it was irrelevant that there was

---

[90]   Commission *Discussion Paper, supra* n. 14, p. 60.
[91]   Ibid., p. 56.
[92]   *Article 82 Guidance, supra* n. 6, p. 18.
[93]   Commission *Discussion Paper, supra* n. 14, pp. 56 and 60 and Case T–201/04 *EC Microsoft*, paras 917–919 and 921–922.
[94]   Commission Notice, Guidelines on the application of Article 81 of the EC Treaty to technology transfer agreements [2004] OJ C101/2 (the TTA Guidelines), para. 191.
[95]   *Jefferson Parish Hospital District No. 2 v Hyde*, 466 U.S. 2, 104 S.Ct. 1551, 80 L.Ed.2d 2 (1984) at 21 and *Eastman Kodak Co. v Image Technical Services, Inc. et al.*, 504 U.S. 451 (1992) at 462.
[96]   Case T–201/04 *EC Microsoft*, para. 918.
[97]   *Article 82 Guidance, supra* n. 6, p. 18.
[98]   Case T–201/04 *EC Microsoft*, paras 22–29 and 917, 920–933.
[99]   *Article 82 Guidance, supra* n. 6, p. 18, *EC Microsoft* Commission Decision, para. 803 and Case T–201/04 *EC Microsoft*, para. 918.

no separate demand for Windows without WMP.[100] That was not to say that the CFI ignored the consumer demand for operating systems with streaming media players; nevertheless it found that others could supply the media player equally well to Microsoft for the original equipment manufacturer (OEM) to combine with the operating system on the ready-to-use PC.[101] The CFI also noted that there was in fact a demand for an operating system without a media player from a particular customer group, namely companies not wanting employees to use computers for non-work-related purposes.[102] Defining a relevant market based on particular customer groups had been applied before in *United Brands*[103] and was then highly criticised.[104] The Commission stated in its Commission Notice on the Relevant Market that defining a market around a particular customer group is only possible if the group could be subject to price discrimination.[105] The CFI did not comment upon this finding further. Yet the question remains, were the Commission and the CFI correct in the finding of two products in *EC Microsoft*?

Returning to the mobile phone example, a mobile phone with a camera cannot be said to be substitutable by a mobile phone and a separate camera, as the former bundle offers special integrated features that the two products separately cannot compete with, such as the ability to send the pictures taken immediately to another mobile.[106] In comparison, purchasing an operating system and later a media player is more easily substitutable by an operating system with an integrated media player. The end result is largely the same. Similarly, consumers would be no worse off – price aside – purchasing cartridges and nails separately rather than purchasing them together in Hilti's bundle. Following this indicator of a stand-alone product neither Hilti's PAFS nor Microsoft's Windows with WMP could be said to be fully integrated.[107]

Art and McCurdy are, however, of a slightly different opinion.[108] While

---

[100]   Case T–201/04 *EC Microsoft*, paras 921–922 and 943.

[101]   Ibid., paras 922–923.

[102]   Ibid., para. 924.

[103]   Case 27/76 *United Brands*, para. 31.

[104]   Jones and Sufrin, *EC Competition Law, Text, Cases and Materials* (3rd edn, Oxford University Press, Oxford, 2007), p. 356.

[105]   Commission Notice on the Definition of the Relevant Market, *supra* n. 29, para. 43.

[106]   Dolmans and Graf, *supra* n. 27, p. 227 – although this was correct at the time of writing in 2007 – now in 2009, digital cameras come with Bluetooth!

[107]   Ibid., p. 228.

[108]   Art, Jean-Yves and McCurdy, Gregory, 'The European Commission's Media Player Remedy in its Microsoft Decision: Compulsory Code Removal Despite the Absence of Tying or Foreclosure' (2004) 25 *European Competition Law Review*, 694. Note both are lawyers for Microsoft.

agreeing that *Hilti* and *Tetra Pak II* dealt with physically distinct products,[109] they argue that Windows operating system's integration with WMP
in *EC Microsoft* requires the codes of the products to work together to enable
the computer to play a CD or download and play a file from the internet.[110]
Art and McCurdy assert that although there are separate markets for both
operating systems and media players, Windows with WMP is technologically integrated; because if the media related code files were taken out of
Windows neither product would work on its own. In comparison, nails and
milk cartons can be sold and used separately from nail guns and filing
machines.[111]

Although this argument is valid to some extent, it does not take into
account the fact that you can download media players onto the operating
system later and Microsoft has indeed managed to produce a workable
version of Windows (Windows Vista N) without a media player. Art and
McCurdy's argument is applicable only to Microsoft's Windows because of
the way it has been set up – but not to operating systems in general. The
Commission assessed this in *EC Microsoft* finding that although there are
other operating systems that integrate a media player functionality, they use
third party media players and allow these to be removed.[112]

Evans *et al.* disagree with this comment and, like Art and McCurdy, argue
in favour of Microsoft that the type of integration Microsoft has applied to
Windows and WMP is also done by Apple, Sun and Linux on their operating systems.[113] Evans *et al.* continue by arguing that restricting the tie
between Windows and WMP will simply hinder European consumers from
benefiting from an integrated software product as well as stifle innovation in
software markets.[114] In particular, integration has become an important
driver of innovation in technology industries because it provides efficiencies
to developers and benefits to consumers, but also has important static effects
on both the demand and supply sides of the industry. Application software

---

[109]   This classification is also referred to by the Commission: *EC Microsoft*
Commission Decision, para. 841 and adopted by several scholars: see for instance
Evans, David S. and Padilla, A. Jorge, 'Tying Under Article 82EC and the Microsoft
Decision: A Comment of Dolmans and Graf' (2004) 27 *World Competition: Law and
Economics Review* 503–512, p. 503.
[110]   Art and McCurdy, *supra* n. 108, pp. 697–698.
[111]   Ibid., p. 698.
[112]   *EC Microsoft* Commission Decision, paras 821–824.
[113]   Evans, D.S., Padilla, A.J., and Polo, M., 'Tying in Platform Software: Reasons
for a *Rule-of-Reason* Standard in European Competition Law' (2002) 25 *World
Competition*, 509, p. 513.
[114]   Ibid.

depends on the software integrated with the operating system.[115] This argument is in line with the one raised by the Court of Appeals in *US Microsoft III*.[116]

In *EC Microsoft*, Microsoft argued that consumers expected to be able to play CDs and DVDs as well as download film and music from the Internet when purchasing a PC.[117] The Commission rejected this argument by noting that consumers would not stop purchasing Windows if there was no WMP, but rather they would get the OEMs to install a third party media player instead,[118] thereby suggesting that Microsoft could tie a media player to Windows; it just could not be WMP.[119] The CFI agreed with this attitude.[120] In 1995, Microsoft had done just that with RealPlayer, however the Commission argued that one abuse (tying Windows with WMP) could not be defended by another (tying RealPlayer to Windows).[121] The Commission's arguments and the CFI's agreement are flawed and do not take into account what is actually happening within the market. They seem to miss the point Microsoft was making that media players are an expected function of all operating systems. Interestingly, Hellström, who was part of the Commission's legal team in the CFI *EC Microsoft* case, has commented that the intervention by the Commission, and therefore the remedy, came too late to stop the media player market from tipping.[122] This comment explains to a certain degree why the Commission approach does not fully match the reality of the current software market.

In comparison, as noted above, when purchasing a mobile phone or any other technologically advanced product today, most consumers would choose the most up-to-date version, which is the one that comes with the most novel technology, and that includes a camera phone. This means that you can sell a camera without a phone, but you cannot sell a phone without a camera! In other words, a separate market for the tied product will often continue to exist even when the bundle has become part of commercial usage. This is also the case in respect of operating systems and media players. You can sell the media player on its own (albeit you need an operating system to use it),[123] but consumer demand requires a media player with an operating system.

---

[115]   Ibid., p. 513.

[116]   *US v Microsoft Corp.(US Microsoft III)*, 253 F.3d 34 (D.C. Cir. 2001), at 95.

[117]   *EC Microsoft* Commission Decision, para. 809.

[118]   Ibid.

[119]   Ibid., para. 822.

[120]   Case T–201/04 *EC Microsoft*, paras 921–923, 943.

[121]   *EC Microsoft* Commission Decision, para. 818.

[122]   Hellström, Per, speech given at 'The Microsoft Judgment' conference, 25 September 2007, BIICL, London.

[123]   Compare with *EC Microsoft* Commission Decision, para. 811; Case T-201/04 *EC Microsoft*, para. 939 and Dolmans and Graf, *supra* n. 27, p. 228.

Dolmans and Graf argue that there are several levels in the market, which dictate whether a component is a component of a product or a separate product. At one end of the market, as for instance with cars, the car is perceived as one product, whereas at another level, that of spare parts and after-sales services, individual car parts are separate products.[124] In other words, products relate to each other on different levels, and therefore even though there may be a demand for a component of a product on its own, it does not necessarily mean that it has been tied to the main product in an abusive manner.[125]

2.1.3.   Tying in 'aftermarkets' and consumables

In certain circumstances, the relationship between the tying and the tied product is complementary. The tied product will be a component, spare part, after-sales service, or consumable of the tying product. Examples of these tied markets are printers and ink cartridges,[126] cars and after-sales services, and vacuum cleaners and vacuum cleaner bags.

Both *Hilti* and *Tetra Pak II* involved tying of consumables. In *Hilti*, the nails constituted the aftermarket and in *Tetra Pak II* it was the cartons for the machines. Where the tied product market is seen as an aftermarket, there is a greater fear of customers being locked in to the product, meaning that once a product is purchased on the primary market, the customer either is forced to purchase after-sales service (for instance when buying a car) or will also have to buy spare parts from the same product manufacturer to maintain the quality and occasionally the value of the product. This was what the EC Courts feared would happen in *Hilti* if the market were left to its own devices.[127]

Nevertheless, it has been seen as rather controversial that companies can be found dominant in their own aftermarkets. It has been argued that a company will not be able to exploit its customers in the secondary market through higher prices if the company does not have a dominant position in the primary market.[128] The Commission does not seem to have taken these aspects into consideration in all its cases. For instance in *Digital*[129] the Commission found

---

[124]   Ibid.

[125]   Others agree with this point: Evans and Padilla, *supra* n. 109, p. 9.

[126]   There have been a few complaints to the Commission regarding this: *Pilikan v Kyocera, XXYth Report on Competition Policy* (1995) part 87, and Commission Decision *Info-Lab/Ricoh*, Case IV/36431 on photocopy machines and toner cartridges.

[127]   See Price, Diana R., 'Abuse of a dominant position – the tale of nails, milk cartons and TV guides' (1990) 11 *European Competition Law Review* 80, p. 85.

[128]   *XXVth Report on Competition Policy* (1995), para. 86; Williams, M., 'Sega, Nintendo and Aftermarket Power: The Monopolies and Mergers Commission Report on Video Games' (1995) 2 5 *European Competition Law Review* 310, p. 313; and Jones and Sufrin, *supra* n. 104, p. 425.

[129]   *Digital, XXVIIth Report on Competition Policy* (1997) and Commission Press

that Digital was dominant in the service market for both hardware and software of its own products despite the fact that the primary market for computer systems was extremely competitive. Such a conclusion is puzzling, given the intense competition on the primary market, but can perhaps be put down to the difference between the price of the primary market product and the transparency of the price and the proportion of prices on the aftermarket.[130]

Case law shows that tying of aftermarkets is seen as equally abusive, and therefore it offers no separate or different assessment of such tying. In other words, the current approach to tying in aftermarkets in EC competition law is strict and inflexible.

2.1.4. Concluding remarks on two separate products
The above analysis makes it clear that there are grounds for asking whether the Commission in particular with respect to *EC Microsoft* has intervened to stop a perfectly natural progression of product integration and innovation in a particular market, which just happened to be led by a powerful company. On the other hand, other aspects of the analysis indicate that the Commission was perhaps correct in its verdict that Microsoft had in fact engaged in illegal tying. However, it is apparent that its analysis is inconsistent and sets some very high standards of proof.

To conclude: the separate consumer demand test is flawed when it comes to more complex and technologically advanced products as it cannot distinguish between product and component/spare part of a product. More alarming is that what can also be deduced from the above case law analysis is that the question which seems relevant to the Commission and the EC Courts is not actually whether there are one or two products, but rather whether there are any independent manufacturers of the tied components. This is irrespective of any IP rights and irrespective of product innovation, market development and stand-alone demand for the tying product. In comparison, in US case law the focus has also been upon consumer demand, but important consideration has been given to the specific market conditions.[131] Although in *EC Microsoft*, the Commission in its anti-competitive effects assessment looked at the specific market conditions, it did not consider these, or rather it dismissed these in rela-

---

Release IP/97/868 (http://europa.eu.int/rapid/pressReleasesAction.do?reference=IP/97/868&format=HTML&aged=1&language=EN&guiLanguage=en).

[130] Andrews, P., 'Aftermarket Power in the Computer Service Market: The Digital Undertaking' (1998) 19 *European Competition Law Review* 176, pp. 177 and 181.

[131] See Chapter 4, Section 2.2.1 and *US Microsoft III* and *United States v Jerrold Electronics Corp.*, 187 F. Supp. 545, 560 (E.D. Pa. 1960), aff'd *per curiam*, 365 U.S. 567 (1961).

tion to the separate product assessment. The CFI permitted the argument of technological integration as an objective justification, but disregarded it as relevant for the separate product assessment.

The problem with the current consumer demand test is that it does not give special consideration to technological integration and particular market conditions, but merely looks at the demand of the tied product market, and this in turn risks stifling innovation.[132] Moreover, the Commission and the CFI did not consider Microsoft's argument of integration until the objective justifications and it here rejected such arguments due to the previous finding of two separate markets and a consumer demand for the WMP, the tied product.[133] It is clear that such an approach is unacceptable if innovation and product development are to be encouraged. The CFI further made it clear that only in the absence of independent demand for the tied product would technological integrated products be seen as one.[134] Although this will be discussed further below, it is important to highlight at this point, because the current approach stifles product development, and the consumer demand test and the relevant market definition are largely to blame. The weight granted to objective justifications is too small in comparison to these tests to allow beneficial technological integrated products to go free of Article 82.

2.2.   Establishing Dominance in the Tying Product Market

Article 82 does not condemn the possession of a dominant position, but merely the abuse of such position. In principle, Article 82 applies *only* when the company is already dominant.[135] Article 82 cannot be applied to companies which act in an abusive manner to *gain* a dominant position.[136] Consequently, case law has developed a low threshold for dominance, thereby allowing the Commission to catch those companies that behave in an abusive manner to *gain* (more) market power.[137]

As noted in Chapter 2, tying is a common business activity among non-dominant companies, as the tie will mostly benefit consumers, and effective competition in the tying product market should ensure that there will remain

---

[132]   Cases C–333/94 P *Tetra Pak II*, para. 36, T–30/89 *Hilti*, para. 67, and T–201/04 *EC Microsoft*, para. 927.
[133]   Ibid., paras 23–29, 917.
[134]   Ibid., para. 918.
[135]   Case 322/81 *Michelin I*, para. 70.
[136]   O'Donoghue and Padilla, *supra* n. 9, p. 107.
[137]   The ECJ has held that market shares exceeding 50 per cent are in themselves evidence of a dominant position: Case 62/86 *AKZO Chemie BV v Commission* [1991] ECR I-3359, para. 60.

realistic alternatives.[138] Therefore, it will normally suffice to establish dominance in the tying product market.[139]

The reason for requiring market power in the tying product market was highlighted by the Commission in *Hilti*:

> The ability to carry out its illegal policies stems from its power on the markets for Hilti-compatible cartridge strips and nail guns (where its market position is strongest and the barriers to entry are highest) and aims at reinforcing its dominance on the Hilti-compatible nail market (where it is potentially more vulnerable to new competition).[140]

The statement corresponds with the Commission's point of view in the TTBER and the TTA Guidelines where tying is seen as likely to create anti-competitive effects only if the licensor holds 'a significant degree of market power' in the tying product market.[141]

In *Hilti*, Hilti was found dominant, holding 70 per cent of the cartridge strips market and 55 per cent of the market for nail guns.[142] In *Tetra Pak II* Tetra Pak had a 90 per cent market share in both the aseptic machine and cartons markets.[143] In *EC Microsoft* Microsoft held a 90 per cent market share in the operating system for personal computers market, in other words, a near-monopoly position.[144]

The ECJ has made a distinction between dominance and super-dominance.[145] The finding of super-dominance will place an additional burden

---

[138] Dolmans and Graf, *supra* n. 27, p. 226.

[139] Cases C–333/94 *Tetra Pak II*, paras 25 and 30–31, and T–201/04 *EC Microsoft*, para. 842; *Hilti* is a deviation from this. Hilti was found dominant in both the tying product market and the tied product market, the nail market; in fact its market share was higher in the tied product market (70–80 per cent): Case T–30/89 *Hilti*, paras 87, 89–92 and 94.

[140] *Hilti* Commission Decision, para. 74.

[141] Tying is not included as one of the hardcore restrictions: TTBER *supra* n. 18, Article 4. The legal effect of this is that tying is fully exempted below a market share threshold of 20 per cent in the case of agreements between competitors and 30 per cent in the case of agreements between non-competitors in respect of technology transfer agreements: the TTA Guidelines, *supra* n. 94, paras 192–193.

[142] Case T–30/89 *Hilti*, para. 85 and *Hilti* Commission Decision, para. 70.

[143] *Tetra Pak II*, Commission Decision, para. 104.

[144] *EC Microsoft* Commission Decision, para. 432.

[145] See Cases C–333/94 P, *Tetra Pak II*; C–395/96 P *Compagnie Maritime Belge Transports SA v Commission* [2000] ECR I–1365, [2000] 4 CMLR 1076, para.114 et seq. The term was only applied in AG Fennelly's opinion (p. 137); however, the CFI judgment endorsed AG Fennelly's conclusion. Appeldoorn, J., 'He Who Spareth his Rod, Hateth His Son? Microsoft, Super-dominance and Article 82 EC' (2005) 26 *European Competition Law Review* 653, p. 653. See also Jones and Sufrin, *supra* n. 104, p. 436 and O'Donoghue and Padilla, *supra* n. 9, p. 166.

upon the company, forcing it to compete less vigorously. This essentially means that there are three levels in the dominance assessment: non-dominance, dominance and super-dominance.[146] In the three tying cases discussed above it is clear that the companies involved all held significant market power or super-dominance. This could be a coincidence as there are so few cases in the area. With only 55 per cent market share in the tying product (nail gun) market, *Hilti* is the case concerned with the lowest market share.[147]

On the other hand, it could be significant that the EC Courts and the Commission have taken a similar approach/understanding to that found in the US, namely that tying is unlikely to cause harm unless the company holds significant market power in the tying product market.[148] This conclusion fits well with the findings in Chapter 2 and the Commission's approach in the TTBER. However, neither the EC Courts nor the Commission have made any clear statement in this respect. The ECJ has indicated that certain conduct may be abusive when performed by a company with significant market power; however, when undertaken by a company with mere dominance it may not constitute abuse.[149] As a result, there is limited assurance in respect of the level of dominance required in tying cases, and trends from Article 82 case law generally suggest a different, stricter approach towards finding dominance, so that even when a company has as little as a 39.7 per cent market share it will be found dominant.[150] This leaves companies with little legal certainty and a presumption that tying is illegal once some level of market power has been achieved, thereby placing greater emphasis on the requirement for dominance in comparison to the requirement of abuse. The previous tying case law leaves room for a more lenient interpretation in favour of a significant market power threshold instead of the lower dominance threshold, and therefore this weighing can (and needs to) be re-adjusted.

2.2.1.   Conclusion on dominance

It is fair to say that companies can currently feel safe from interference by the competition authorities only if they have no market power. In market share terms, that would be below 30 per cent as equivalent to the safe harbours allowed for under the TTBER. This is in fact false security for Article 82 as, in principle, it comes into play only if the company is dominant; that is if it

---

146   Appeldoorn, *supra* n. 145, p. 653.
147   Note, however, Hilti had 70–80 per cent of the tied product (nail) market.
148   See Chapter 4 and Chapter 5.
149   Case C–395/96 P *Compagnie Maritime Belge*, para. 114 et seq., and Faull and Nikpay, *supra* n. 14, pp. 336–337.
150   See Case T–219/99 *British Airways Commission* [2003] ECR II–5917, [2004] 4 CMLR 1008, para. 211.

holds more than (approximately) a 40 per cent market share.[151] In the US, *Jefferson Parish* made it clear that a 30 per cent market share was insufficient to establish market power; hence in this respect the two jurisdictions are in concurrence.[152] EC case law indicates that 'super dominance' has been the norm in tying cases. Yet the emphasis in the cases was upon dominance; once found dominant the companies were easily found guilty of abuse as well. This is in sharp contrast to recent case law development in the US where the market power requirement is *no more* than one of the conditions necessary to finding a tying arrangement illegal.[153]

2.3. Abuse

Article 82 has been elusive in its wording and examples to provide a concrete definition of abuse. Case law has evolved around a definition of exclusionary abuse given by the ECJ in *Hoffmann-La Roche*.[154] The definition contains three conditions. First, abuse is an objective concept – its presence is based on facts and not upon whether the company intended to behave in an abusive manner.[155] That said, intent can play a role in assessing the effects of the abuse, and as such can be used as evidence against the company.[156] Second, it is practices which differ from those of normal competition or competition on the merits which are abusive.[157] In other words, the analysis must look to the form of the conduct, but naturally requires some definition of what competition on the merits is.[158] The difficulties in distinguishing between abuse and competition on the merits have resulted in an aggressive application of Article 82,[159]

---

[151]   The *Article 82 Guidance, supra* n. 6, p. 7.

[152]   *Jefferson Parish Hospital District No. 2 v Hyde*, 466 U.S. 2, 104 S.Ct. 1551, 80 L.Ed.2d 2 (1984) at 7, 13–14 and 27 and Chapter 4, Section 2.2.2.

[153]   *US Microsoft III* at 84 and Chapter 4, Section 2.2.2.

[154]   Case 85/76 *Hoffmann-La Roche*, para. 91.

[155]   Faull and Nikpay, *supra* n. 14, p. 349; Vickers, *supra* n. 7, p. 417; Case 85/76 *Hoffmann-La Roche*, para. 91.

[156]   In Case 62/86 *AKZO*, evidence of intent was applied to prove the abuse; see also Case T–201/04 *EC Microsoft*, para. 911, and Bellis, Jean-François, 'The Microsoft Judgment' Conference, 25 September 2007, BIICL, London.

[157]   Case 85/76 *Hoffmann-La Roche*, para. 91 and Cases T–65/98 *Van den Bergh Foods Ltd. v Commission* [2003] ECR II–4653, [2004] 4 CMLR 1, para. 157; T–228/97 *Irish Sugar Plc v Commission* [1999] ECR II–2969, para. 111; and T–203/01 *Michelin v Commission (Michelin II)* [2003] ECR II–4071, para. 97.

[158]   Cases 62/86 *AKZO*; C–333/94 P *Tetra Pak II* paras 39–44, *Digital* Commission Decision, Case T–340/04 *France Télécom SA v Commission* [2007] ECR II–107, and O'Donoghue and Padilla, *supra* n. 9, p. 176.

[159]   *EC Microsoft* Commission Decision, n. 877, where the Commission argued that even common practice within an industry was not necessarily normal competition

which does not sufficiently take into account efficiencies and consumer benefits which certain conduct may generate, but instead applies a formalistic approach, which does not correspond to economic theory.[160] Moreover, this condition has led to the development of the special responsibility that dominant companies are said to hold; a responsibility which increases with the level of market power held.[161] The special responsibility, originating from the principle of fairness,[162] implies a positive obligation upon the dominant company and that a form of relationship exists between the dominant company and the market in which it operates.[163] Third, the conduct should have anti-competitive effects which restrict both current competition and potential growth of competition.[164] This last condition of the exclusionary effect will essentially require different assessments based on the form of abuse investigated.

Forcing the purchase of two products together, or coercion, is at the very heart of the tying analysis. The assessment of coercion or abuse is therefore closely related to the anti-competitive effects assessment. If there is no evidence of anti-competitive effects there is no coercion, and vice versa. Article 82(d) specifically requires that the contracts are 'subject to acceptance of supplementary obligations'. In case law, coercion has been interpreted as consumers being unable to choose to obtain the tying product without the tied product.[165]

---

when undertaken by a dominant company. See also Cases T–201/04 *EC Microsoft*, para. 942 and C–333/94 P *Tetra Pak II*, para. 37.

[160]   Rousseva, E., 'Abuse of Dominant Position Defences, Objective Justification and Article 82 EC in the Era of Modernisation', Chapter 10 in Amato, G. and Ehlermann, C-D. (eds), *EC Competition Law, A Critical Assessment* (Hart Publishing, Oxford, 2007), p. 380; Kallaugher, J. and Sher, B., 'Rebates revisited: Anti-competitive Effects and Exclusionary Abuse under Article 82' (2004) 25 *European Competition Law Review* 263, and OECD Policy Brief, 'What is Competition on the Merits?', June 2006 (www.oecd.org/dataoecd/10/27/37082099.pdf).

[161]   Cases 322/81 *Michelin I* para. 57 and C–333/94 P *Tetra Pak II*, para. 24.

[162]   Amato, Giuliano, *The Antitrust and the Bounds of Power, the Dilemma of Liberal Democracy in the History of the Market* (Hart Publishing, Oxford, 1997), p. 67. The principle of fairness originates from the theory of ordoliberalism. 'The ordoliberalists believed that economic competition would provide the basis for the society they envisioned, but only where law could create and maintain the conditions under which competition could function properly'; Gerber, *supra* n. 1, p. 241.

[163]   Bavasso, A., 'The Role of Intent under Article 82 EC: From "Flushing the Turkeys" to "Spotting Lionesses in Regents Park"' (2005) 26 *European Competition Law Review*, 616, pp. 617–618 and Al-Dabbah, Maher M., 'Conduct, Dominance and Abuse in "Market Relationship": Analysis of Some Conceptual Issues under Article 82 EC' (2000) 21 *European Competition Law Review*, 45, p. 46. Note though that the concept was left out of the Commission *Discussion Paper*, *supra* n. 14, and was not mentioned by the CFI in *EC Microsoft* either. However, it features in the *Article 82 Guidance*, *supra* n. 6, pp. 4 and 5.

[164]   Case 85/76 *Hoffmann-La Roche*, para. 91.

[165]   Case T–201/04 *EC Microsoft*, para. 944.

In both *Hilti* and *Tetra Pak II*, the key question for determining coercion was whether customers were realistically hindered from purchasing the products separately and were thereby excluded from the choice of buying the products individually.[166] Consumer choice, or lack of it, is therefore clearly an antitrust concern, as not only does coercion affect consumers in limiting their choice of products, but it can leave suppliers at a competitive disadvantage as well,[167] and ultimately lead to exclusion of competitors.[168]

In *EC Microsoft*, the same conclusion was come to despite the fact that the WMP was not forced upon the customer, but given away for free as an integrated part of Windows.[169] The Commission, supported by the CFI in *EC Microsoft*, seemed to go one step further than previous case law, arguing that coercion occurred as a result of the fact that consumers were restricted in the initial process of choosing a media player with an operating system. In reality, however, after the choice of operating system is made, the consumer can still opt for other media players and is not obliged to make use of it; hence, there were in fact limits to the coercion.[170]

Case law has shown that the forced purchase or coercion can appear in various ways, either directly or indirectly. For instance, it can take the form of a contractual clause, as in *Tetra Pak II*, or a refusal to supply the products separately, as in *Hilti*.[171] Here the refusal was disguised by a sales technique of claiming that the individual products were part of a package. Another type of tying could be to eliminate guarantees, as was found in *Novo Nordisk*, where Novo Nordisk was found to have abused its dominant position by not upholding guarantees and disclaiming liability for malfunctioning when its pen products were used with other compatible products from other manufacturers.[172] Equally in *Hilti*, Hilti also refused to honour its guarantees where the customer had not used Hilti's nails but those of a third party in Hilti's guns.[173] Another form, which is more controversial, is product integration such as was claimed in *EC Microsoft*.[174]

---

[166] *Hilti* Commission Decision, para 75 and Case T–83/91 *Tetra Pak II*, para. 137.

[167] *EC Microsoft* Commission Decision, para. 833.

[168] *Hilti* Commission Decision, para 75; Case T–201/04 *EC Microsoft*, para. 945; and Monti, G., *EC Competition Law* (Cambridge University Press, Cambridge, 2007), p. 191.

[169] *EC Microsoft* Commission Decision, paras 800, 830 and 833.

[170] Case T–201/04 *EC Microsoft*, para. 960 and Page, W. and Lopatka, J., 'The Dubious Search for "Integration" in the *Microsoft* Trial' (1998–99) 31 *Connecticut Law Review* 1251, p. 1270.

[171] *Hilti* Commission Decision, para. 75.

[172] *Novo Nordisk, XXVIth Report on Competition Policy* (1996), pp. 142–143, (www.europa.eu.int/comm/competition/publications/26en2_en.pdf (26 January 2006)).

[173] *Hilti* Commission Decision, para. 79.

[174] *EC Microsoft* Commission Decision, paras 800–813 and Case T–201/04 *EC Microsoft*, para. 963.

A more common form is financial tying, such as was seen in *Digital*.[175] Digital applied other forms of tying, proving that combinations of the above practices can also occur. Digital offered combined hardware and software packages, which the Commission found to be a deliberate policy to restrict competitors in the market for hardware maintenance services for Digital computers systems.[176] Additionally, Digital applied fidelity discounts to customers provided they did not obtain supplies from other service providers during the duration of the contract with Digital.[177]

Dolmans and Graf argue that the Commission's view is that the mere identification of financial tying could be enough to prove abuse of a dominant position.[178] However, *Digital* can also be read another way. Undoubtedly, the Commission will consider intervening if exceptional price discounts are given when two products or services are purchased together. Nevertheless, the Commission will only do so if it is the discount that makes it uneconomical for the customer to go elsewhere and where the discount is not proportionate to the costs.[179] Andrews agrees and notes that this means that the Commission sees both non-predatory and non-discriminatory discounts as abusive in conjunction with tying, unless costs savings or other benefits passed on to the consumers can justify the use of them.[180]

The Commission's policy is also in line with that of the ECJ. In *Michelin I*, which was about discounts on tyres, the ECJ held that for the discount to be abusive it had to be investigated:

> whether, in providing an advantage not based on any economic services justifying it, the discount tends to remove or restrict the buyer's freedom to choose his sources of supply, to bar competition from access to the market, to apply dissimilar conditions to equivalent transactions with other trading parties or to strengthen the dominant position by distorting competition.[181]

This confirms the two points illustrated in *Digital*: firstly that the discount is not based on costs or economies of scale, that is to say a quantity discount,[182] and secondly the effect the discount will have on the customer's ability to choose. Although *Michelin I* did not revolve around tying, it emphasised the

---

175   *Digital* and Commission Press Release IP/97/868, *supra* n. 129.
176   *Digital*, pp. 130–131.
177   Ibid.
178   Dolmans and Graf, *supra* n. 27, p. 231.
179   *Digital*, p. 131.
180   Andrews, *supra* n. 130, pp. 178–179.
181   Case 322/81 *Michelin I*, para. 73.
182   Case C–163/99 *Portugal v Commission: Landing Fees at Portuguese Airports* [2001] ECR I–2613, [2002] 4 CMLR 1319, para. 49.

attitude taken by both the ECJ and the Commission and directly correlated with *Hilti*. The approach was confirmed in *British Airways*, where the ECJ held that anti-competitive effects can be counterbalanced by 'economic justifications', which must not only be efficient but also benefit consumers.[183]

Therefore, for a tie not to be abusive under Article 82 the customers must have a meaningful commercial choice to purchase the products separately.[184] This choice will disappear if the tying product is sold at the same price as the tied package.[185] The irony is that this is the result of the Commission's remedy in *EC Microsoft*; so although the Commission perhaps was correct in finding that Microsoft had abused its dominant position by integrating the WMP with Windows or giving it away for free alongside Windows, its own remedy has still left the consumers with no real choice.[186]

In a footnote, the Commission argued that 'WMP's "price" could arguably be deemed to be "hidden" in the overall price for the bundle of Windows and WMP'.[187] It is, however, well known that competitors also offer media players free of charge and the cost of a single media player is insignificant. The CFI agreed with the Commission on this point and found it irrelevant for the conditions under Article 82(d) to be fulfilled that the WMP was obtained for free and that consumers were not obliged to use the WMP and were able to obtain other media players (for free) to use with Windows.[188] In principle, the CFI is correct in rejecting these points as important for the coercion analysis. Nevertheless, these facts are indeed relevant to the assessment of foreclosure and therefore for the tying assessment. The CFI's concern was that the OEMs were forced to take the bundle and thereby indirectly passed on the coercion to consumers.[189] This concern has left some commentators to argue that the case was in fact not about the tie, but Microsoft's distribution system.[190]

Finally, *intent* seems to have played an important role in the finding of abuse. Questions have been raised as to why Microsoft employed the tying strategy. Some have argued that the tying strategy was designed to move Microsoft closer to the entertainment industry.[191] The Internet is increasingly attracting music, video and games businesses, and for Microsoft, being the

---

[183]   Case C–95/04 *British Airways v Commission* [2007] ECR I–2331, [2007] 4 CMLR 22, para. 86.
[184]   Dolmans and Graf, *supra* n. 27, p. 231.
[185]   Ibid.
[186]   *EC Microsoft* Commission Decision, para. 831.
[187]   Ibid., n. 971 – see also Case T–201/04 *EC Microsoft*, para. 968.
[188]   Ibid., paras 968–970.
[189]   Ibid., para. 962.
[190]   Art and McCurdy, *supra* n. 108, p. 700.
[191]   Rushe, Dominic, 'Microsoft braced for big fines by EU', *The Sunday Times*, 21 March 2004.

world's largest software company, entering into this industry is an opportunity not to be missed.[192] Both the software market and the entertainment industry are part of the group labelled high technology industries, a fact which has not gone unnoticed by the Commission. It stated, as an explanation for the tie being anti-competitive, that Microsoft had a clear incentive for seeking a stronghold in the media player market and found that the media player market was a 'strategic gateway' to related markets.[193] Gaining dominance in this market would give Microsoft a competitive advantage in markets such as content encoding software, format licensing, wireless information device software, digital rights management solutions, and online music delivery markets.[194] Allowing the tie, the Commission argued, would make Microsoft the gatekeeper of the media content market earning profits from each sale in the market.[195]

Another argument has been that the tie was applied as a defence by Microsoft to protect its near monopoly in the operating system. A media player used in conjunction with middleware such as Java could substitute for an operating system, become a 'fully-fledged' platform in itself and thereby be a significant threat to Windows.[196] Needless to say, Microsoft has disputed this wholeheartedly.[197] Nevertheless, economists agree with this theory and note that it is a likely strategy for a company producing consecutive products or new generation products.[198] It thus seems a plausible reason behind Microsoft's tie. However, for this strategy to work the dominant company must be fully committed to the tie and not sell the products separately,[199] and this is not the case with Microsoft. One could speculate whether this is because Microsoft is already so powerful that it does not need to be fully committed to obtain some advantage from the tie, or perhaps Microsoft is unaware of the economic theories surrounding such strategies.

The CFI relied heavily upon one particular email between a Microsoft executive and Bill Gates as evidence for Microsoft's anti-competitive intent behind its integration of Windows and WMP. In fact it shaped most of its ruling around the email, which conveyed that Microsoft could apply technological integration to push RealPlayer out of the market by 'reposition [the] streaming media battle from NetShow vs. Real to Windows vs. Real' and

---

[192]   Ibid.
[193]   *EC Microsoft* Commission Decision, para. 975.
[194]   Ibid.
[195]   Ibid., para. 1067.
[196]   Ibid., paras 971–972.
[197]   Ibid., para. 971.
[198]   See Chapter 2, Section 5.1.2.
[199]   Nalebuff, *supra* n. 44, pp. 58–59.

'follow the [Internet Explorer] strategy wherever appropriate'.[200] It could be argued that the CFI applied intent to circumvent the real problem of the case, whether the technological integration of WMP with Windows created efficiencies which outweighed any anti-competitive effects the tie had upon the media player market.[201]

The intent behind the tying arrangement in *EC Microsoft* played a significant, if not the determining, role in finding coercion and thereby finding the tie abusive. With such great emphasis upon this email and the CFI's limited interest in identifying clear evidence of anti-competitive effects, it is tempting to argue that the intent demonstrated in the email was the only real evidence in *EC Microsoft* for Microsoft's behaviour being abusive. This application of intent goes beyond previous rulings by the ECJ and is an unfortunate use of intent.[202]

2.3.1. Conclusion on abuse

In conclusion, coercion can be established either directly or indirectly, intentionally or unintentionally. As a matter of fact, the conclusion as to whether coercion exists is based on the previous assessment of dominance and whether there are two separate products or an integrated single product. If there are two separate products and these are sold together by the dominant company, then the Commission and the EC Courts have been quick to find coercion due to lack of consumer choice, and thus abuse. Only in *EC Microsoft* did intent play a significant role in the outcome of the case, a development which should not be followed lightly.

2.4. Anti-competitive Effects – Foreclosure of the Tied Product Market

The anti-competitive effect analysis remains the most controversial aspect of abuse, as the EC Courts have not been consistent in their requirement for this element. The ECJ has at times required proof of exclusion of *all* competition,[203] whereas recently the CFI in *British Airways* held that:

---

[200]   Case T–201/04 *EC Microsoft*, para. 911, and also 937, 1352, and Bellis, *supra* n. 156.

[201]   Harchuck, K.E., 'Microsoft IV: The Dangers to Innovation posed by the Irresponsible Application of a Rule of Reason Analysis to Product Design Claims' 97 (2002) *Northwestern University Law Review* 395, p. 431–433.

[202]   Case 85/76 *Hoffmann-La Roche*, para. 91; Faull and Nikpay, *supra* n. 14, p. 349; and Harchuck, *supra* n. 201, pp. 431–433.

[203]   Cases 6 & 7/73 *Istituto Chemioterapico Italiano SpA & Commercial Solvents Corp. v. Commission* [1974] ECR 223, [1974] 1 CMLR 309, para. 25; C–7/97 *Oscar Bronner*, para. 41 and Faull and Nikpay, *supra* n. 14, p. 150.

[it] is sufficient in that respect to demonstrate that the abusive conduct . . . tends to restrict competition, or in other words, that the conduct is capable of having, or likely to have such an effect.[204]

It thereby set a very low standard of proof, which appears to have been applied in recent tying case law.

There are at least three reasons for tying being viewed as anti-competitive. First, it can exclude competitors, work as an entry deterrent, and drive customers away from third party products in the tied product market;[205] second, it limits consumer choice, which eventually can lead to exploitative abuse;[206] and, third, it may affect two or more markets, as by tying products together the dominant company attempts to leverage its market power from one market into another.[207] Hence, although tying is directed towards customers, it has an indirect effect of excluding competitors or deterring potential competitors.[208]

There are two elements to establishing whether a tie has created anti-competitive effects. Firstly, there is a need to prove that the tie creates (future) anti-competitive effects and, secondly, that the effects are sufficiently serious to distort competition.[209] The latter element is the one which has received the most attention, just like in the US. The former element is almost taken for granted in case law. It is assumed that all ties will create some anti-competitive effect; the question is how much. Nevertheless, there may be room for improvement here as economists are of the opinion that a tie will not always be purely anti-competitive; it can also be pro-competitive. It is thus inadequate merely to focus upon the anti-competitive effects without considering pro-competitive elements.

In case law, a less sophisticated test seems to have been applied by the Commission to assess whether the tie in question had adverse effects on

---

204   Case T–219/99 *British Airways*, para. 293, judgment upheld on appeal in Case C–95/04 *British Airways*.
205   *Article 82 Guidance, supra* n. 6, p. 17, Harbord, D. and Hoehn, T., 'Barriers to Entry and Exit in European Competition Policy' (1994) 14 *International Review of Law and Economics* 411, p. 418 and O'Donoghue and Padilla, *supra* n. 9, pp. 486–488.
206   *Hilti* Commission Decision, para 75; Case T–201/04 *EC Microsoft*, para. 945 and Monti (2007), *supra* n. 168, p. 191.
207   *Hilti*, Commission Decision, para. 74.
208   *EC Microsoft* Commission Decision, para. 794; see also Commission *Discussion Paper, supra* n. 14, p. 54; and Anderman, *supra* n. 15, p. 148.
209   In its *Discussion Paper*, the Commission requires the customer group that competitors cannot reach due to the dominant company's behaviour to be identified and assessed as to whether that group constitutes a sufficient part of the market: Commission *Discussion Paper, supra* n. 14, pp. 56–57.

competition. For instance in *Napier Brown/British Sugar*,[210] where British Sugar had tied its sale of sugar to the delivery transport, the Commission did not even consider whether the tying had foreclosed or had any anti-competitive effect upon the transport market. The fact that British Sugar had reserved the separate sugar delivery market for itself was evidence enough to establish the requirement of anti-competitive effect.[211]

Likewise, in *Hilti* the Commission merely concluded about the tying conduct 'these policies all have the object or effect of excluding independent nail makers who may threaten the dominant position Hilti holds'.[212] This form of assumption has also been applied by the ECJ as shown in *Télémarketing*, where the ECJ only stated that Télémarketing reserved to itself 'an ancillary activity which might be carried out by another undertaking as part of its activities on a neighbouring but separate market, with *the possibility* of eliminating all competition from such undertaking'[213] (emphasis added). In these cases, no factual evidence as such was required to prove anti-competitive effects. It was sufficient to show that the conduct was likely to have anti-competitive effects amounting to a risk of excluding competitors from the market.

In *EC Microsoft*, the Commission took a different view; it found that Microsoft's tying had the *potential* to risk foreclosure in the future.[214] The Commission almost uniquely provided a rather extensive analysis of anti-competitive effects and concluded by offering factual evidence for the risk of foreclosure.[215]

It is questionable whether such an extensive analysis of anti-competitive effects would have occurred if the Commission had had a strong case against Microsoft.[216] As the Commission itself explained, despite the tying of the media player with the operating system there remained an option for consumers to obtain third party media players through downloading on the internet.[217] The Commission therefore rightly noted that foreclosure cannot be assumed and an investigation is required.[218] The Commission thereby admitted that *EC*

---

[210]   *Napier Brown/British Sugar*, Commission Decision 88/519/ EEC [1988] OJ L284/41.

[211]   Ibid., para. 46.

[212]   *Hilti* Commission Decision, para 75 and Ahlborn, Evans and Padilla (2004), *supra* n. 16, p. 36.

[213]   Case 311/84 *Centre Belge d'Etudes du Marché-Télémarketing v Compagnie Luxembourgeoise de Télédiffusion SA and Information Publicité Benelux SA* [1985] ECR 3261, [1986] 2 CMLR 558, para 27.

[214]   *EC Microsoft* Commission Decision, paras 842–954.

[215]   Ibid.

[216]   Ibid., paras 841–954.

[217]   Ibid., para. 841.

[218]   Ibid.

Microsoft is different both from previous case law and from contractual tying. This is important because the Commission thus introduced a more thorough assessment to the special surroundings of the *EC Microsoft* case.

Due to the complexity of many technologically advanced products and high-technology industries, it would be precarious not to extend Article 82 to technological integration or fast-moving industries, although allowing for such extension requires a much greater responsibility for ensuring the right balance between competition regulation and innovation. Therefore, the Commission was correct in engaging in a more scrupulous analysis. Whether it was the technological integration or special market condition which triggered this more thorough anticompetitive effects assessment remains a little hazy. However, technological integration can most likely be ruled out because the Commission rejected the fact that WMP was an integrated part of Windows. The Commission's approach in *EC Microsoft* does not compare with its strict approach in the *Article 82 Guidance*.[219] This leads to the conclusion that the Commission chose this more thorough anti-competitive assessment because of the market conditions; in other words, there were other distribution options available. Such an approach corresponds better with the approach in *US Microsoft III*.[220] If this is a correct interpretation, the Commission has taken a sensible and responsive approach for which it should be commended although it has given no clear indication of when this more perceptive approach should be applied. Moreover, the Commission has not clarified whether it is only when alternative options are available for obtaining the tied product that a more thorough anti-competitive assessment is called for, or whether other market conditions can trigger this more sensitive approach. Microsoft, on the other hand, did not see the extensive anti-competitive effects analysis as a positive feature; rather it challenged it before the CFI as being a further condition in the tying assessment and asserted that in reaching its conclusion of foreclosure the Commission had 'relied on a new and highly speculative theory'.[221]

Art and McCurdy criticised the Commission for having expanded the tying doctrine in a 'dangerously broad manner' to find an infringement of Article 82.[222] Although they are correct in questioning the Commission's handling of *EC Microsoft*, the problem does not lie in the expansion of the tying doctrine, as they argue, but rather the fact that the Commission has not quite managed to achieve the right balance between competition regulation and innovation.

---

[219]   The *Article 82 Guidance*, *supra* n. 6, p. 18.
[220]   *US Microsoft III* at 89–90.
[221]   Case T–201/04 *EC Microsoft*, paras 846 and 989.
[222]   Art and McCurdy, *supra* n. 108, p. 695.

Eilmansberger agrees that the Commission chose the wrong tools to assess the case, although his conclusion is based on the assumption of a distinction in the definitions of tying and bundling, arguing that the Commission assessed Microsoft as a tying case, whereas in reality it was a bundling case.[223] 'Bundling is here understood to mean the inclusion or addition of one or more components in a given product or service to increase the appeal of the latter, irrespective of whether this leads to a higher price or not.'[224] Bundling has an additional function to tying which is to add value to a product by adding a complementary product, and for this reason bundling should be seen as competition on the merits and not as simple exclusionary behaviour.[225] Nevertheless, Eilmansberger concludes that in certain circumstances even bundling can be anti-competitive, and thus the Commission was correct in finding Microsoft's behaviour abusive.[226] Eilmansberger's theory is interesting; however it automatically adds an additional assessment criterion to tying conduct: the assessment of the type of tying arrangement, an aspect of tying which has so far been ignored by the Commission. Eilmansberger's theory aligns itself with the approach in *US Microsoft II*, where the Court of Appeals decided to assess the actual product design.[227] Whether this is referred to as bundling or technological integration, the fact remains that the Commission rejected this line of reasoning when it identified two separate products, and no real distinction was thus allowed between contractual tying and technological integration.

In comparison, the CFI did not enter into a debate as to whether Microsoft's tying was similar to previous case law; it merely noted that in previous tying cases foreclosure effects had been considered, and thus the Commission was correct in assessing this aspect, even though it hinted that the Commission could have got away with a less thorough assessment of foreclosure effects.[228] Second, the CFI relied, as already mentioned, upon one particular email, which provided evidence of intent to leverage by Microsoft; that the integration of the WMP with Windows was designed to make WMP more competitive with RealPlayer, the then dominant media player in the market.[229] The CFI further emphasised the consequences of such conduct by noting that the

---

[223]   Eilmansberger, Thomas, 'How to Distinguish Good from Bad Competition Under Article 82EC: In Search of Clearer and more Coherent Standards for Anti-competitive Abuses' (2005) 42 *CMLRev* 177, p. 154.

[224]   Ibid., p. 153.

[225]   Ibid. p. 154.

[226]   Ibid.

[227]   *United States v Microsoft Corp.*, 147 F.3d 935 (D.C. Cir. 1998) (*US Microsoft II*).

[228]   Case T–201/04 *EC Microsoft*, paras 857–859.

[229]   Ibid., paras 911, 937 and 1352.

integration makes WMP the platform of choice for complementary content applications, which in turn creates a risk of foreclosing competition on the media player market, as well as affecting neighbouring markets, such as media encoding and management software.[230]

The flaw in both the Commission's and the CFI's findings of foreclosure was not the application of the tying condition, nor the extensive analysis, but the fact that the focus was mainly on the distribution of the WMP with Windows, not on the media player market itself. Microsoft argued that the Commission should in its evidence show more than the mere fact that Microsoft distributed Windows together with WMP to prove harm to competition,[231] because there were other ways of reaching customers than via Microsoft's OEM pre-installation, such as downloading from the Internet. Likewise, Microsoft noted that it is mainly in PCs for home use that WMP is pre-installed with Windows, whereas OEM pre-installation is not the way to reach business users of software who account for more than 55 per cent of all new PCs sold.[232]

The Commission and the CFI found that due to the tie there was a disincentive among OEMs to ship PCs with an additional media player, as it would take up hard-drive capacity and would require additional costs in respect of training of staff, customer support and testing, with the only result that customers were unlikely to pay a higher price for the additional (and duplicate functionality) software.[233] The OEMs were thus restricted in offering to consumers the PC package – that is software and hardware combination – which they found the most attractive.[234]

Alternative distribution channels were assessed in the case such as downloading from the Internet and bundling of the media player with other software or Internet access services or separate sale of the media player at retail level.[235] The Commission argued that consumers found downloading more difficult by comparison with OEM pre-installation and that it required fast/broadband connections to the Internet, which denied less frequent users of the Internet access to downloading.[236] Microsoft disputed this, arguing that constant development of the media player market allowed for quicker and better downloads.[237]

---

[230]   Ibid., para. 857.
[231]   *EC Microsoft* Commission Decision, para. 839.
[232]   Ibid., para. 850.
[233]   Ibid., paras 851–852 and 857; Case T–201/04 *EC Microsoft*, para. 1044.
[234]   Ibid., para. 1046; compare with Evans, D. S., Nichols, A.L., and Schmalensee, R. 'United States v Microsoft: Did Consumers Win?' (2005) 1 *Journal of Competition Law and Economics* 497.
[235]   *EC Microsoft* Commission Decision, paras 858–871 and 872–876.
[236]   Ibid., paras 866–868 and Case T–201/04 *EC Microsoft*, para. 1050.
[237]   *EC Microsoft* Commission Decision, para. 868.

Another concern for the Commission was the network effects created by the content providers and software developers targeting their products to specific media players. The Commission feared that in presence of network effects downloading of competing products of equal quality to WMP was restricted.[238] The Commission found downloading less advantageous and unlikely to offset the negative effects of the tie between the WMP and Windows.[239] The CFI concurred and noted that more than one media player on an operating system would confuse the consumer.[240] Both thus concluded that downloading was not compatible with OEM pre-installation because it was less efficient and the downloading was often of a second media player, making it less attractive to use.[241]

In principle, whether a media player is combined with the operating system as the first or the second media player does not really matter; what matters is access to the operating system and thereby access to the consumer. An operating system–media player bundle need not be a one-to-one tie, but can easily be one-to-two or -three.[242] Access to an operating system can be obtained in three ways: (1) via pre-installation on the computer together with the operating system, (2) via individual sale in shops, and (3) via download from the Internet. All options require that the operating system manufacturer actually allows operability. One way of further promoting a media player and making it more attractive to consumers, and distinguishing it from other media players, is by having content providers and software developers target the specific media player or have a unique product connected with the media player, as Apple has done with its iPod and iTunes.[243]

The Commission's and the CFI's rejection of alternative distribution channels does not sit easily with an objective of ensuring innovation. The software market is fast moving and customers are also quick to adapt. Hence, Microsoft may have an initial competitive advantage in being able to pre-install its WMP with Windows,[244] but other media players still have significant opportunities within the media player market to reach customers. Interestingly, there has been an increase in media players in the market since Microsoft commenced integration,[245] although this increase perhaps could have been greater if

---

[238] Ibid., para. 861.
[239] Ibid., para. 871.
[240] Case T–201/04 *EC Microsoft*, para. 1045.
[241] Ibid., paras 1049–1050 and *EC Microsoft* Commission Decision, paras 859–860.
[242] See Chapter 2, Section 3.3.1 and Bellis, *supra* n. 156.
[243] *EC Microsoft* Commission Decision, para. 138.
[244] Case T–201/04 *EC Microsoft*, para. 1047.
[245] Ibid., para. 1055.

Microsoft had not integrated.[246] The CFI applied a similar line of reasoning in *British Airways*.[247] However, only the CFI applied this approach by noting that users with pre-installed WMP were less likely to use alternative media players, but in the absence of bundling consumers would be induced to choose from all available media players in the market.[248]

The Commission found that Microsoft's argument of downloading an equivalent alternative to pre-installation contradicted Microsoft's insistence upon its own rights to pre-install.[249] However, the Commission is here forgetting that Microsoft considered that its WMP was part of its operating system and Microsoft was merely attempting to prove that the media player market had not been foreclosed due to the integration. The alternative distribution channels, including downloading, were dismissed because they were second bests, not because the tie foreclosed the market as such (the Commission later concluded that there was a future risk of this or rather that this was already occurring).[250]

The Commission and the CFI were wrong in not giving the alternative distribution channels greater weight in the foreclosure effect assessment. This has been further emphasised by the fact that the remedy has been argued to be useless and too late to save the market, partly due to the development of the market and more people having broadband Internet connections.[251] Microsoft has developed a version of Windows without WMP, for which it has found limited demand.[252]

The Commission found that Microsoft had interfered with the normal competitive process and cycles of innovation by applying the tie of Windows and WMP to create strong network effects and thereby barriers to entry.[253] Microsoft was therefore shielding itself from effective competition and potentially more efficient media player producers.[254] 'Microsoft's conduct affects a market which could be a hotbed for new and exciting products springing forth in a climate of undistorted competition.'[255] The Commission even went so far as to say that the tying of the WMP to Windows was sending warning signals

---

[246]   Marsden, Philip, comment given at 'The Microsoft Judgment' Conference, 25 September 2007, BIICL, London.

[247]   Case T–219/99 *British Airways*, para. 298.

[248]   Case T–201/04 *EC Microsoft*, para. 1041.

[249]   Ibid.

[250]   Art and McCurdy, *supra* n. 108, p. 702 and *EC Microsoft* Commission Decision, para. 923.

[251]   See Bellis, *supra* n. 156, and Hellström, *supra* n. 122.

[252]   Case T–201/04 *EC Microsoft*, para. 943.

[253]   *EC Microsoft* Commission Decision, para. 980.

[254]   Ibid., para. 981.

[255]   Ibid.

to other companies considering innovating in markets which Microsoft might also have an interest in entering.[256] The CFI concurred.[257]

The Commission unsurprisingly found that there was a reasonable likelihood that the tying conduct would lead to the lessening of competition in the near future.[258] The CFI in fact found it unnecessary for the Commission to have embarked upon this analysis of establishing the ubiquity of WMP due to the integration, as it held that the finding of the restriction of distribution was sufficient to fulfil the foreclosure requirement.[259] Not surprisingly, the CFI held that the indirect network effects created by Microsoft stemming from the bundle further enforced the foreclosure effect, giving Microsoft a competitive advantage unrelated to the merits of its product, which also affected neighbouring markets.[260] The CFI therefore concluded that the Commission had sufficiently demonstrated that Microsoft's tie risked the foreclosure of the market. It found 'a reasonable likelihood that tying Windows and [WMP] would lead to a lessening of competition so that the maintenance of an effective competition structure would not be ensured in the foreseeable future'.[261] The statement simply reiterates the Commission's conclusion, but also follows the line taken in *British Airways;*[262] that of a very low standard of proof for anti-competitive effects. As was illustrated above, the foreclosure element of tying cases has been infamously vague. The CFI's conclusion in *EC Microsoft* provides greater clarification of the standard of proof. However, it offers little in respect of a more economic or lenient approach towards tying arrangements. First, the statement makes clear that the anti-competitive effects need not have occurred, but that there is a 'reasonable likelihood' that they will take place in the foreseeable future. Importantly, in comparison to previous case law 'lessening of competition' is sufficient, rather than the likely elimination of all competition. That said, the 'lessening of competition' must still be so significant that it will weaken the effective competitive structure of the market. Yet, case law has shown that the presence of a dominant company will mean that the effective competitive structure of a market is already perceived as weakened.[263] Therefore, the last part of the CFI's conclusion appears to be already fulfilled when a tying company is found dominant in the tied product market.

---

[256] Ibid., para. 983.
[257] Case T–201/04 *EC Microsoft*, para. 1088.
[258] *EC Microsoft* Commission Decision, para. 984.
[259] Case T–201/04 *EC Microsoft*, paras 1058–1059.
[260] Ibid., paras 1060–1069 and 1076.
[261] Ibid., para. 1089.
[262] Case T–219/99 *British Airways*, para. 293.
[263] Case 85/76 *Hoffmann-La Roche*, para. 91.

There seems to be a discrepancy between the Commission's understanding of anti-competitive effects and foreclosure requirements under a tying assessment and the CFI's ruling above. The Commission suggested in its *Discussion Paper* that to fulfil the market distortion foreclosure effect two conditions must be met.[264] First, rival companies are not able to compete with the dominant company for its customers.[265] Second, there must be an assessment of whether the market as a whole can be considered foreclosed.[266] In comparison, the *Article 82 Guidance* merely lists a set of factors which the Commission finds important for identifying 'likely or actual anticompetitive foreclosure'.[267] The Commission's standard of proof is therefore a little unclear. The suggestion for the standard of proof given in its *Discussion Paper* is higher than the one applied by CFI in *EC Microsoft* and, additionally, seems to be in line with Article 82 refusal to supply cases, which have some similarities with tying cases. In these cases, the ECJ referred to the elimination of competition.[268] However, the comments in the *Article 82 Guidance* are more in line with the CFI's lower standard: competitors in the media player market need not be in a position of (almost) leaving the market, but a reduction in their sales is sufficient.

There appears to be no reasonable rationale for the Commission and the CFI to establish a lower standard for tying in comparison to refusal to supply. In fact this is a dangerous route to wander down, in particular because refusal to supply and tying share many common features, and some tying arrangements could be viewed as refusal to supply, as was the case in *Télémarketing*. The introduction of the lessening of competition standard by the CFI is probably inspired by its own low effects standard developed in *British Airways* rather than a well-thought-through notion of distinguishing between refusal to supply cases and tying. It is therefore questionable how much emphasis should be placed upon this new standard for tying.

The terminology, lessening of competition, is in line with the language of the US Clayton Act, Section 3, and to some extent the approach under the current EC Merger Regulation.[269] The US case law on tying indicated that

---

264 Commission *Discussion Paper*, *supra* n. 14, pp. 56–57.
265 Ibid.
266 Ibid., p. 58.
267 *Article 82 Guidance*, *supra* n. 6, pp. 8–10 and 18.
268 See Cases 6 & 7/73 *Commercial Solvents*, para. 25; C–241–242/91 P *Radio Telefis Eireann v Commission (Magill)* [1995] ECR I-743, [1995] 4 CMLR 718, para. 56; C–7/97 *Oscar Bronner*, para. 41; C–418/01 *IMS Health*, paras 38 and 52; compare with Case T–65/98 *Van den Bergh*, para. 160 and 172 where contributing 'significantly to the foreclosure of the relevant market' seemed sufficient.
269 Council Regulation (EC) No 139/2004 of 20 January 2004 on the control of concentrations between undertakings (the EC Merger Regulation) [2004] OJ L24/1.

although the foreclosure effect need not be present, a risk of it is sufficient; the impact though still has to be substantial upon the tied product market.[270] This final requirement does not seem to be so explicit in the CFI's ruling.

2.4.1. Concluding remarks on anti-competitive effects of tying

Case law shows that the anti-competitive effects assessment is indeed a grey area, in particular, due to the lack of certainty as to what needs to be identified and the standard of proof. Case law prior to *EC Microsoft* indicated that a finding that the tying was capable of or likely to lead to foreclosure was sufficient; in other words, no evidence was needed to prove that exclusion of competitors had already taken place. In some ways, this is reasonable if preventative rather than corrective competition regulation is wanted. However, as noted by Ridyard, it is the speculative nature of the anti-competitive effects which is troubling.[271] The risk of false positive errors increases due to the fact that these cases are often initiated by competitors and not customers.[272] Thus, there is a risk of excessively vigorous regulation if interference is based on mere assumptions of foreclosure rather than if a higher standard of proof like reasonable probability of foreclosure is required.[273] The *EC Microsoft* Commission Decision offers an alternative in this respect, as the Commission left no stone unturned in its efforts to support its claim of risk of foreclosure of the tied market. Nevertheless, even this assessment had its flaws. In comparison, in a rigid and very legalistic ruling, the CFI reverted to previous case law, and reiterated the low standard of proof by clarifying that the threshold which the evidence had to meet was a reasonable likelihood of lessening competition in the foreseeable future. The CFI indicated that the Commission had gone beyond this threshold in its decision. In other words, the threshold for the standard of proof was lower than the evidence presented by the Commission. Such a low standard of proof is problematic if tying is not to be treated as illegal *per se*, because it will mean that whether tying is abusive is decided upon the basis of dominance and not anti-competitive effects. This goes against the very wording and core of Article 82, which specifically notes that it is the abuse which is illegal, not the dominance in itself. Case law has given no indication of intentions to treat contractual tying and technological integration differently. In comparison, the *Article 82 Guidance* signifies a suggestion for a stricter treatment of technological integration. This would be in opposition to the recommendations in Chapter 2.

---

[270] *Jefferson Parish* at 16.
[271] Ridyard, Derek, 'Tying and Bundling – Cause for Complaint?' [2005] ECLT – European Competition Law Review 316, p. 317.
[272] Ibid.
[273] Kallaugher and Sher, *supra* n. 160, p. 284.

2.5. Objective Justifications

Unlike Article 81 with its subsection (3), Article 82 does not contain an express exemption clause, although in principle the requirement of dominance establishes an automatic exemption. Case law has demonstrated the necessity for an exemption and granted exemptions where there was an objective justification for the behaviour. In *Télémarketing* this was referred to as any 'technical or commercial requirements relating to the nature'[274] of the product, which could justify the behaviour.[275]

In *Masterfoods*,[276] AG Cosmas introduced the principle of proportionality: '[a] company which holds a dominant position is not entitled to bring about disproportionate restrictions to free competition, even if the goals sought are wholly legitimate . . .'.[277] AG Cosmas' definition of objective justification sets a high standard of proof for the defending dominant company.

In respect of tying, the objective justification analysis permits an opportunity for the dominant company to defend the tying arrangement as efficient or otherwise welfare enhancing.[278] In EC case law, we have only three major cases dealing with this issue, and all three failed to fulfil the conditions attached to the objective justification.

In *Hilti*, Hilti argued that it was concerned with the reliability, operation and safety of PAFS, in particular in respect of the use of third party products with any of the components of the PAFS.[279] The Commission, however, rejected this argument for three reasons. First, Hilti had provided no evidence of previously having taken health and safety issues into consideration for instance, through notification to customers and competitors.[280] Moreover, the Commission found no evidence of any accidents attributable to the application of third party nails with Hilti's nail gun.[281] Second, the Commission held that the existing EC safety standards were sufficient safeguards.[282] This argument has now been re-enforced by the Commission in its *Article 82 Guidance*, which states '[it] is not the task of a dominant company to take steps on its

---

274   Case C–311/84 *Télémarketing*, para. 26.
275   The objective justification test is linked to the concept of proportionality developed under other areas of EU law, such as free movement of goods, services and workers.
276   Case C–344/98 *Masterfoods Ltd v. HB Ice Cream Ltd* [2000] ECR I–11369, [2001] 4 CMLR 449.
277   AG Cosmas' Opinion in Case C–344/98 *Masterfoods*, para. A101.
278   O'Donoghue and Padilla, *supra* n. 9, p. 230.
279   *Hilti* Commission Decision, para. 87.
280   Ibid., paras 89.4–90.
281   Ibid., para. 93.
282   Ibid., para. 92.

own initiative to exclude products which it regards, rightly or wrongly, as dangerous or inferior to its own products'.[283] Third, the Commission argued that Hilti could have applied other measures less restrictive than tying to uphold high health and safety standards. In other words, the Commission believed Hilti was motivated by concerns other than safety.[284]

In *Tetra Pak II*, Tetra Pak also relied on the health and safety argument as well as the protection of its reputation and product reliability. The CFI dismissed the argument by referring to *Hilti*. It held that when there were less restrictive measures available, which would not infringe Tetra Pak's IP right, it was not for Tetra Pak to impose measures on its own initiative even if it was for health and safety reasons and the protection of its reputation.[285] Tetra Pak also argued that certain of the contested clauses in its licensing agreements, where it reserved exclusive rights over modification and maintenance and had IP rights over any improvements or modifications to the equipment, were justified on grounds of security and efficiency.[286] In response, the CFI held that the tied-sale clauses in Tetra Pak's contracts clearly went beyond their 'ostensible purpose' and were unreasonable for the protection of public health.[287]

In both *Hilti* and *Tetra Pak II* the accused companies attempted to defend their tying arrangements by claiming that there was complementary usage between the products. It was indisputable that without compatible nails the cartridge strips would be useless. This implies that there may not have been a distinct demand for the cartridge strips separate from the nails. Yet, it does not remove a separate demand for nails.[288] The CFI's judgment in *Hilti* made it clear that if there is a demand by the consumers for the tied product to obtain it from an alternative source, then the dominant company is obliged to offer consumers a realistic choice to do so regardless of whether the products are complements or otherwise linked.[289] The ECJ endorsed this view in *Tetra Pak II*, when Tetra Pak claimed in defence of its tying arrangements for cartons and filling machines that there was a natural link between the two products, and the tied sales were in accordance with commercial usage.[290] The ECJ rejected this interpretation because of the finding of independent manufacturers in the tied cartons market, which were not active in the cartons machines market.[291]

---

[283] *Article 82 Guidance, supra* n. 6, p. 12.
[284] *Hilti* Commission Decision, para. 90.
[285] Case T–83/91 *Tetra Pak II*, paras 138–139.
[286] Ibid., para. 131.
[287] Ibid., para. 140.
[288] Case T–30/89 *Hilti*, para. 68 and Dolmans and Graf, *supra* n. 27, p. 229.
[289] Ibid.
[290] Case C–333/94 *Tetra Pak II*, paras 34 and 35.
[291] Ibid., para. 36 and Case T–83/91 *Tetra Pak II*, para. 82.

The ECJ concluded by stating that the list of abusive practices in Article 82 is not exhaustive.[292] Therefore even if there is a natural link between products or the tied sale is in accordance with commercial usage the tie may still be abusive unless it can otherwise objectively justified.[293]

Finally, in *EC Microsoft*, Microsoft listed two types of efficiency defence in the Commission Decision: efficiency related to distribution and efficiencies related to WMP as a platform for content and applications.[294] Firstly, Microsoft argued that having a set of default options in a personal computer reduced the transaction costs for consumers by reducing time and confusion.[295] The Commission rejected this as it felt that Microsoft had failed to take into account that the media players pre-installed by the OEMs needed not be a Microsoft product.[296] The arguments highlighted by the Commission seem to be contrary to the arguments it made in respect of downloading; moreover they appear also to confirm Art and McCurdy's suspicion that the real purpose of the Commission's investigation was an attempt to get Microsoft to make use of other companies' media players. This is further confirmed by the CFI's statement that it was concerned with the restriction Microsoft's tying brought upon the OEMs and not the actual product integration.[297]

Secondly, Microsoft argued in the Commission Decision that transaction costs were reduced due to the tied sale of the two products as it saved resources spent on maintaining a separate distribution system for the second product and it ensured that the efficiencies gained were passed on to customers.[298] The Commission acknowledged this argument as valid and would have accepted it; however, it felt that 'such savings cannot possibly outweigh the distortion of competition in this case. This is because distribution costs in software licensing are insignificant'.[299]

It is rather surprising that the Commission has acknowledged this as a valid defence. Although this is also a persuasive argument made by economists,[300] if one accepts this, thinking of a hypothetical situation where a dominant company bundles its product together with that from another market where it does not hold a dominant position, or perhaps a new product altogether, the fact that the dominant company can save significant costs by applying one

---

292   Case C–333/94 *Tetra Pak II*, para. 37.
293   Ibid.
294   *EC Microsoft* Commission Decision, paras 955–970.
295   Ibid., para. 956.
296   Ibid., paras 956–957.
297   Case T–201/04 *EC Microsoft*, paras 1041–1049, and 1149.
298   *EC Microsoft* Commission Decision, para. 958.
299   Ibid.
300   Chapter 2, Section 4.1 and Salinger, Michael A., 'Graphical Analysis of Bundling' (1995) 68 *The Journal of Business* 85, p. 98.

distribution network to two (or more) products should not be the only ground for allowing the tie. Regardless of the fact that this would make sense economically, it would mean that companies like United Brands, which was found to be dominant in the market for bananas, could in principle be allowed to tie bananas together with oranges provided that the savings were sufficiently large. Ridyard (an economist) agrees with this point, noting:

> [an] approach that says bundling is acceptable by dominant firms as long as they can provide an 'objective justification' does not hold the solution . . . either. [As] the assumption behind this approach is that it may be possible for the dominant firm to cancel out the harm it has done through bundling if it can find an offsetting efficiency advantage.[301]

The fact that the Commission has accepted the defence therefore raises the question not only of why it has done so, but also how the objective justification and efficiency defences should be argued. The quest may not therefore be for one particular justification but for a cluster of several that together reach the threshold.

Microsoft did not rely on the transaction cost argument again in the CFI ruling. It argued defence of complementary products: that the WMP was an integrated part of Windows.[302] The Commission chose to ignore this argument as it had previously rejected this under the two separate products assessment.[303] As the analysis above shows, it is not entirely clear that the Commission or the CFI was correct in stating that there were two separate products. This conclusion is thrown further into doubt when one reads the comments made in the *Article 82 Guidance*, which states that technological integration by combining two products in a new innovative way for the benefit of consumers will be considered as an efficiency defence.[304] In its *Discussion Paper*,[305] the Commission had originally required evidence of the combination of the two products to be so efficient that the whole industry

---

[301]   Ridyard (2005), *supra* n. 271, p. 317 compare with Kühn, Kai-Uwe, Stillman, Robert and Caffarra, Christina, 'Economic Theories of Bundling and their Policy Implications in Abuse Cases: An Assessment in Light of the Microsoft Case' (2005) 1 *European Competition Journal* 85.

[302]   *EC Microsoft* Commission Decision, paras 800 and 955.

[303]   Ibid., para. 955.

[304]   *Article 82 Guidance*, *supra* n. 6, p. 20.

[305]   For comments on the *Discussion Paper's* assessment of efficiency defences see Ahlborn, C., Denicolò, V., Geradin, D. and Padilla, A.J., 'DG Comp's Discussion Paper on Article 82: Implications of the Proposed Framework and Antitrust Rules for Dynamically Competitive Industries' 31 March 2006 (http://papers.ssrn.com/sol3/papers.cfm?abstract_id=894466), p. 39.

would in the future offer the integrated product.[306] This has now been removed from the *Article 82 Guidance* together with the comment that technological integration was more likely to fulfil the conditions of the efficiency defence than contractual tying.[307]

Microsoft claimed that WMP integrated into Windows would lead to superior technical product performance.[308] It argued that Windows would be weakened if the two products were not integrated, as they supported each other.[309] The Commission rejected this argument, holding that Windows had previously operated without WMP and therefore could easily do so again,[310] and the argument could not be accepted as objective justifications because '[the] existence of such inter-dependencies would be the result of a deliberate choice by Microsoft'.[311]

In comparison, the CFI held that it was not the actual bundle that was the problem, but the fact that Microsoft did not also offer a version of Windows without WMP.[312] It should be recalled that the main argument for the WMP not being an integrated part of Windows was that there were several independent manufactures of media players in the market.[313] The Commission's argument and the CFI's ruling fail to take into consideration the dynamics of high-technology industries and the trend within those to innovate via integration.[314] Consequently, not permitting this form of innovation risks stifling innovation in high-technology markets. However, the remedy applied – the mixed bundling – ensures to some extent that innovation can continue.

Microsoft argued that Windows operates faster with WMP integrated, but never provided the evidence for this.[315] It argued that the integration guaranteed media functionality on which software developers and internet site creators relied.[316] Removing the media functionality would be to the detriment

---

306   Ibid., p. 60, n. 125.

307   Commission *Discussion Paper, supra* n. 14, p. 60.

308   Case T–201/04 *EC Microsoft*, paras 1146–1147 and *EC Microsoft* Commission Decision, para. 962.

309   Case T–201/04 *EC Microsoft*, paras 1146–1147; *EC Microsoft* Commission Decision, paras 1019 and 1026 and Krim J., 'EU Orders Microsoft to Modify Windows', *Washington Post*, 23 December 2004.

310   *EC Microsoft* Commission Decision, para. 1019.

311   Ibid., para. 1027 and Case T–201/04 *EC Microsoft*, paras 1151–1152.

312   Case T–201/04 *EC Microsoft*, para. 1149.

313   Ibid., para. 927 and *EC Microsoft* Commission Decision, para. 804.

314   Evans, Padilla and Pollo, *supra* n. 113, p. 74; Ahlborn, Denicolò, Geradin and Padilla, *supra* n. 305, p. 41; and Monti, G., 'Article 82 EC and New Economy Markets' Chapter 2 in Graham, C. and Smith, F. (eds), *Competition, Regulation and the New Economy* (Hart Publishing, Oxford, 2004), pp. 48 and 50.

315   Case T–201/04 *EC Microsoft*, para. 1160.

316   *EC Microsoft* Commission Decision, paras 962–963 and 967.

of consumers, software developers and internet site creators as well as degrade and fragment the operating system.[317] The Commission argued that the efficiencies highlighted by Microsoft were all achievable with cooperation between operating systems and media players regardless of whether or not the bundle consisted only of Microsoft components.[318] Strangely, Microsoft never attempted to demonstrate beyond this that the integration created some technical efficiencies which could not be recreated by merely tying an operating system with a media player, despite the Commission's suggestion of this in its Decision and the Court of Appeals' assessment of this in the US.[319] When prodded by the CFI, Microsoft argued that numerous software developers chose to apply the media player functionality in Windows and this created technical efficiencies due to 'uniform integration'.[320] The CFI rejected this argument due to lack of evidence, stating that application of Microsoft's software did not demonstrate technical efficiencies,[321] and found that Microsoft was:

> not entitled to rely on the fact that bundling ensures uniform presence of media player functionality in Windows . . . [That fact] is precisely one of the main reasons why . . . the bundling led to the foreclosure of competing media players from the market. Although the uniform presence . . . may have advantages for [software developers and internet site creators], that cannot suffice to offset the anti-competitive effects of the tying issue.[322]

Although it can rightly be argued that both the Commission and the CFI placed less weight upon the objective justification arguments and evidence than they put upon the evidence for anti-competitive effects, Microsoft clearly made a blunder by not arguing the technical efficiencies justification better, or perhaps it simply could not provide the evidence for such claims.

Microsoft argued strongly in the Commission Decision that it was common practice in the industry to pre-install the operating system with a media player, and if Microsoft was not allowed to do so it was put at a competitive disadvantage compared to the other operating system manufacturers.[323] It repeated this argument before the CFI, claiming that the Commission had departed from the conditions laid down in Article 82(d).[324] Not surprisingly this argument was

---

317 Case T–201/04 *EC Microsoft*, paras 1146–1147.
318 *EC Microsoft* Commission Decision, paras 964–966.
319 Ibid., para. 962; Case T–201/04 *EC Microsoft*, para. 1159; *US Microsoft II* at footnote 12.
320 Case T–201/04 *EC Microsoft*, para. 1161.
321 Ibid., para. 1161.
322 Ibid., para. 1151, and *EC Microsoft* Commission Decision, para. 968.
323 Ibid., para. 959.
324 Case T–201/04 *EC Microsoft*, paras 844–846.

rejected by both the Commission and the CFI. Firstly, the Commission noted that the other operating system manufacturers allowed for the pre-installation of their operating systems with *a* media player, unlike Microsoft, which only allowed for the pre-installation of Windows with WMP.[325] Secondly, the Commission referred to *Tetra Pak II* where the ECJ held that despite the conduct being in accordance with commercial usage it might be an abuse when undertaken by a dominant company.[326] The CFI concurred, holding that the Commission had followed the correct conditions for tying abuse by applying Article 82 in its entirety instead of focusing upon the wording of Article 82(d).[327] The CFI admitted that there was a link between the two products, yet it found that they could be distinguished for economic and commercial reasons under competition rules.[328] The CFI also noted that it was difficult to talk about commercial usage when 95 per cent of the market was dominated by Microsoft.[329]

With this ruling the CFI has set some very high standards in respect of objective justification. It has shown that it is indeed willing to assess these, but the onus is upon the allegedly dominant company to provide adequate arguments and evidence. Microsoft did not do so. The CFI indicated that in order to accept the bundle as a technological integration, the tying company must show that the bundle offers technical efficiencies leading to superior technical product performance, which must be said to be a positive development towards accepting that protection of innovation is a valid objective justification. Again, Microsoft only argued that Windows operated faster because of the integration; the CFI found this insufficient to fulfil the superior technology condition. The requirement is similar to the approach taken by the Court of Appeals in *US Microsoft II*,[330] albeit that the CFI did not discuss the issue further. Importantly, the discussion of technological integration arrived at too late a stage in the assessment, namely under the objective justification defence. If it had been brought in already under the two separate product assessment, it would have played a greater and more realistic role and would still form part of the defence of the tying arrangement.

2.5.1. Concluding remarks on objective justifications

The above analysis has demonstrated that although the Commission and the EC Courts are willing to assess objective justifications, all bar one have so far

---

[325] *EC Microsoft* Commission Decision, para. 959.
[326] Ibid., para. 961.
[327] Case T–201/04 *EC Microsoft*, paras 859–861.
[328] Ibid., para. 939.
[329] Ibid., para. 940.
[330] *US Microsoft II* at 949.

been rejected, leaving a very high threshold to be reached for dominant undertakings wishing to tie their products. The high threshold, as we shall see, is consistent with the one found under US antitrust law, but not consistent with the economic thinking highlighted in Chapter 2. The high threshold ultimately means that once the company is found dominant, its tying practice becomes illegal *per se*, or, as O'Donoghue and Padilla put it, accepting objective justifications in theory but not allowing for them in practice is the equivalent to no objective justifications at all.[331] This conclusion should be seen in the light of the findings under section 2.4 above where it was identified that the anti-competitive effects threshold was very low; so low in fact that the mere likelihood of foreclosure in the future was sufficient. Therefore, as things currently stand, whether tying is abusive will depend on two elements: (1) definition of the relevant markets and (2) the finding of dominance in the tying product market, leaving the anti-competitive effects and objective justification analysis meaningless. With the low threshold for the former and the high threshold for the latter, anti-competitive effects will always outweigh pro-competitive effects in this approach.

Case law makes it clear that the dominant company must provide evidence of its defences. The mere proof that the conduct is common practice in the industry or commercial usage is not enough.[332] *EC Microsoft* makes it clear that the evidence the Commission and the EC Courts are looking for is of an economic nature. The Commission accepted the reduction in distribution costs as a true efficiency defence and one that it did allow Microsoft to rely on; however, the problem here was clearly that the costs (that is the actual amount of money) saved were not sufficient in comparison to the harm caused.

The CFI clarified that for a claim of a technological integration to be successful, the tying company must show that the integration will lead to superior technical product performance. Microsoft did not manage to demonstrate this, nor did it seriously attempt to do so. Ahlborn *et al.* comment that the small number of tying cases makes it problematic to ascertain whether the objective justification threshold is unreasonably high or whether in the few cases available for assessment the justifications merely were not supported by the facts.[333] It is most likely a combination of both. In particular, *EC Microsoft* sheds light on how difficult it really is to reach the standards expected and the more specific nature of proof required. The conclusion must be that the following defences, whether based on objective necessity or efficiency, can in principle be accepted:

---

[331] O'Donoghue and Padilla, *supra* n. 9, p. 232.
[332] See comments above; Case T–83/91 *Tetra Pak II*, para. 137 and Case T–201/04 *EC Microsoft*, para. 940.
[333] Ahlborn, Evans and Padilla (2004), *supra* n. 16, p. 314.

- health and safety
- product quality (superior technical performance)
- reduction in costs.

As regards health and safety, the main requirement is that the concern is genuine and that the company has taken steps to ensure high safety standards by informing competitors and customers as well as being able to show that the tie is the least restrictive measure to uphold safeguards.[334] Given previous case law and the Commission's comments in its *Article 82 Guidance* it is perhaps unlikely that the threshold for health and safety will ever be reached, rendering it in practice unreachable.

In respect of product quality, the threshold requirement is a product that has technical qualities superior to others in the market, and that this is obtained, amongst other things, via the tie. The Commission in *EC Microsoft* had held that the main evidence necessary is that of the indispensability of the tie; in other words, the superiority cannot be achieved without the tie.[335] However, the CFI remained silent on the matter, thereby leaving the question of standard unanswered. The Commission's indispensability requirement would lead to a very high standard of proof, which essentially could eliminate technological integration as an objective justification.[336] A reduction in costs as an efficiency defence was accepted in *EC Microsoft*. Although, the evidence was good enough, the actual gain was too small as a counterweight to the anti-competitive effects also resulting from the tie.[337] Finally, the key to reaching the utopian standard of proof may be to establish a cluster of defences which together can create the high level required. Needless to say, claims alone are not sufficient (economic or material): evidence must be provided as well.

3. REMEDIES

Regulation 1/2003[338] lays down the powers of the Commission for the enforcement of the competition rules, including the framework for reme-

---

[334]   The company can probably not require higher levels of safety standards than the ones afforded by law: *Hilti* Commission Decision, para. 92 and *Article 82 Guidance, supra* n. 6, p. 12.

[335]   *EC Microsoft* Commission Decision, para. 963.

[336]   Apon, J., 'Cases Against Microsoft: Similar Cases, Different Remedies' (2007) 28 *European Competition Law Review* 327, p. 331.

[337]   *EC Microsoft* Commission Decision, para. 958.

[338]   Council Regulation (EC) No 1/2003 of 16 December 2002 on the implementation of the rules on competition laid down in Articles 81 and 82 of the Treaty [2003] OJ L1/1.

dies.[339] The purpose of the remedy is not only to punish the company for the abusive conduct, but also to promote a means to restore effective competition in the market, thereby putting a stop to the abuse. Additionally it works as a deterrent for other companies. Hence, the most general remedies enforced by the Commission have consisted of demanding that the dominant company terminate its conduct.[340] However, in certain circumstances the remedy has been more controversial.[341]

The breadth of the remedial powers the Commission holds is best illustrated in *Tetra Pak II* where the Commission, with the ECJ's approval, restricted future conduct as well, even though the illegal conduct had already been brought to an end.[342] The Commission ordered that 'Tetra Pak shall refrain from repeating or maintaining any act or conduct described in Article 1 and from adopting any measure having equivalent effect',[343] thereby ensuring that similar action by Tetra Pak in the future would automatically be deemed illegal. The same approach was applied in *Hilti* where Hilti was ordered to terminate the infringements it had been found guilty of, and to desist from repeating or continuing the infringements or measures of equivalent effect.[344] The remedy was upheld by both the CFI and ECJ.

The remedies adopted in *EC Microsoft* differ considerably from the ones ordered in *Tetra Pak II* and *Hilti*. The core purpose of the remedy is the termination of the abuse. Previously that would have required Microsoft to stop tying, but Microsoft was allowed to continue to sell Windows with WMP as long as it also offered an unbundled version of Windows, a so-called mixed bundle.[345] It could therefore be argued that the remedy is more lenient than Hilti's and Tetra Pak's, because in effect Microsoft can continue to tie. On the other hand, the remedy may be more invasive because Microsoft was required to take its products apart, or rather develop a new product, Windows Vista N, while ensuring that this unbundled version of Windows would not perform less well than the bundled version.[346]

---

[339]   It replaced the old Council Regulation 17 1 May 2004.
[340]   Regulation 1/2003, *supra* n. 338, Article 7(1).
[341]   See Cases 6 &7/73 *Commercial Solvents*; *Magill TV Guide/ITP, BBC, and RTE*, Re (Commission Decision 89/205) [1989] OJ L78/43, [1989] 4 CMLR 757; Cases C–241–42/91P *Magill*, and *Sealink/B&I Holyhead: Interim Measures* [1992] 5 CMLR 255.
[342]   *Tetra Pak II* Commission Decision, Article 3 and Case C–333/94P *Tetra Pak II*, para. 51.
[343]   *Tetra Pak II* Commission Decision, Article 3.
[344]   *Hilti* Commission Decision, Articles 1(1) and 3.
[345]   *EC Microsoft* Commission Decision, para. 1011.
[346]   Ibid., para. 1012, see also paras 1013–1014.

It is rather surprising that the Commission omitted to make any comments in respect of future Windows versions and the price of the unbundled Windows. From the Commission's perspective, it should be important to ensure that the unbundled version is not sold at an extortionate price and that the price of the unbundled version should, as a bare minimum, reflect costs of production or be reasonable and non-discriminatory. Though such a requirement is in theory the correct approach and also the one used by the Commission and Courts in other cases such as compulsory licensing cases,[347] in practice, however, the cost of adding WMP to Windows is minute. Hence, in principle, if the price should reflect costs, the price difference between the unbundled version of Windows and Windows with WMP will also be minimal. What the remedy proposes is, as mentioned above, a mixed bundle, which again in theory should eliminate the anti-competitive effects and harm to consumers that the tying conduct caused. Where the unbundled and the bundled versions are offered at almost the same price due to costs structure, as would be the case with Microsoft, the effect of the remedy is limited.[348] The remedy in *EC Microsoft* therefore seems to be misleading of the Commission's policies and rather useless to consumers as they will still be compelled to buy the bundle. However, this should not be seen in a negative light, but rather understood as technological progress; for instance nobody complains about air conditioning now being part of a standard package for a car or a camera being a function of a mobile phone.

Throughout the entire case, Microsoft claimed that Windows with WMP was an integrated product; it therefore argued that the remedy was pointless as there was no significant consumer demand for an unbundled version of Windows.[349] The Commission's counter-argument was that '. . . the user benefits which derive from obtaining a pre-installed bundle of an operating system *and* a media player will therefore not be eliminated whilst the ability to *choose* the media player component of the bundle will be restored'.[350] Subsequently, Microsoft was ordered by the Commission to fulfil the obligations of the remedy within 90 days and forced to have a monitoring trustee assessing whether Microsoft had fulfilled its remedy obligations.[351] The CFI upheld the Commission's finding of abuse and its remedy of providing an unbundled

---

[347]   See Case C–241-42/91P *Magill*, the Commission has also applied this standard in the telecommunication sector: see Commission's Guidelines on the Application of the EC Competition Rules in the Telecommunications Sector [1991] OJ C233/2.

[348]   Case T–201/04 *EC Microsoft*, para. 943, Bellis, *supra* n. 156, and Hellström, *supra* n. 122.

[349]   *EC Microsoft* Commission Decision, para. 1024.

[350]   Ibid., para. 1025.

[351]   Ibid., para. 1017.

version of Windows.[352] However it annulled the request for a monitoring trustee.[353]

Microsoft is now offering unbundled versions of Windows – Windows XP N and Windows Vista N – although these have had limited success.[354] Monti notes in respect of high technology markets that competition law remedies are problematic as they may arrive too late because the market may have corrected itself or the remedies risk stifling innovation, thus harming consumer welfare.[355] The operating systems market and the media player market may not have corrected themselves but certainly developed faster than the competition law enforcers. Bellis, a lawyer working for Microsoft, argued that the remedy was designed to work on the OEM channel distribution, by allowing the OEMs to use an operating system without WMP.[356] They could thereby enter into exclusive agreements with other media player suppliers. This has so far not happened, partly due to downloading being an efficient distribution channel and partly due to the OEMs not wanting to distribute other media players.[357] The remedy is therefore not effective for the purpose it was intended. Furthermore, it does not hinder Microsoft from tying other middleware products to its operating system as long as it continues to offer an unbundled version of Windows. Although from an economic perspective this should ensure the least harm to the tied product market, it is uncertain that it will in reality make much difference or confine Microsoft's influence on middleware markets. Whinston observes that by offering a mixed bundle consumers will be only marginally better off while Microsoft's distribution costs will increase.[358] However, if Microsoft were forced to do this with all its functions of Windows the effect would be much more severe.[359] There was no mention of what Microsoft could or should do in respect of future functionality developments to Windows.

3.1.   Concluding Remarks on Remedies

Having only case law to rely on and no market analysis of what happened in the markets after Hilti, Tetra Pak and now Microsoft were requested to cease

---

352   Case T–201/04 *EC Microsoft*, para. 1229.
353   Ibid., para. 1279.
354   Ibid., para. 943.
355   Monti (2004), *supra* n. 314, p. 23 – see also Hellström, *supra* n. 122.
356   Bellis, *supra* n. 156.
357   Evans, Nichols and Schmalensee, *supra* n. 234, p. 537.
358   Whinston, M.D., 'Exclusivity and Tying in *U.S. v. Microsoft:* What We Know, and Don't Know' (2001) 15 *Journal of Economic Perspectives*, 63, p. 78.
359   Ibid.

their abusive tying, it is difficult accurately to assess whether the remedies were appropriate and re-introduced or re-enforced competition in those markets. Both Hilti and Tetra Pak are still strong players in the nail gun market and aseptic and non-aseptic machine and carton markets respectively, and Microsoft is still dominant in the media player market and, unsurprisingly, the operating systems market. But it is not their dominant position that should be questioned but rather whether their behaviour has been sufficiently curtailed to ensure that other players stand a chance in these markets. Without expert knowledge to confirm this, it appears to be the case for Hilti and Tetra Pak, but it should be highlighted that these markets are considered to be less high-technology and innovative than the computer software industry, and this of course will also affect the success of the remedy. We have yet to see the full impact of the Commission's remedy for Microsoft; however, given the specific market conditions, it is doubtful whether the remedy will greatly influence the market conditions, in particular because consumers now expect all operating systems to arrive with a media player, and content providers which are restricted by costs and therefore not able to write to all media players will choose the media player with the greatest presence in the market to write their content to. This choice has already taken place for many and is unlikely to change in the future. It is beyond the scope of this book thoroughly to assess the remedies and alternatives; however, the above albeit brief analysis suggests that there is indeed a problem in relation to remedies and high-technology industries. The Commission's compromise of requesting mixed bundling of the operating system and the media player to Microsoft demonstrates sensitivity towards innovation, but is perhaps too soft an approach towards abusive behaviour. With Microsoft's track record it will probably not be long before its behaviour is under the Commission's scrutiny again.[360]

4.   ARTICLE 82, TYING AND IP RIGHTS

IP rights will generally fall within the sphere of competition law when they are applied as a leveraging device in refusal to supply/licenses cases and tying cases. As most of the Article 82 tying cases (all bar one – *Napier Brown/British Sugar*) actually involved IP rights, it is necessary to assess the

---

    [360]   See Commission Press Release, 'Antitrust: Commission initiates formal investigations against Microsoft in two cases of suspected abuse of dominant market position', MEMO/08/19, 14 January 2008 and Waters, Richard, 'EU launches new Microsoft antitrust probe', *Financial Times*, 14 January, 2009 (www.ft.com/cms/s/0/2ce90532-c2c1-11dc-b617-0000779fd2ac,_i_email=y.html) and US Department of Justice, Antitrust Division (www.usdoj.gov/atr/cases/ms_index.htm).

relationship between IP rights and Article 82 in greater detail and pose the question: has the presence of IP rights in these tying cases played a decisive role or has their presence merely been incidental? Both *Hilti* and *Tetra Pak II* are illustrations of the Commission's policy in general towards IP right-protected products. The Commission has tended to apply a strict approach to such products in relation to identifying the relevant market, which more often than not was established as the scope of the right itself.[361] Although IP rights in themselves are not presumed to create dominance,[362] they will be strong contributory factors in establishing dominance.[363] Finally in relation to abuse where the application of an IP right is not abusive as long as the exclusion is part of normal exercise,[364] when, however the IP right is applied with a view to leveraging into a downstream market and excluding others from that market, in case law it is not regarded as part of the normal exercise and therefore abusive.

The Commission has in general applied a narrow market definition when establishing the relevant market, and in particular in IP rights cases. The Commission has often defined the relevant market as being along the borders of the national right that the IP right owner holds.[365] It has even gone so far as to establish that there can be a separate market for the spare part of a particular product, even if the product is IP right-protected.[366] This method of reasoning has also been backed up by the EC Courts, which have established that as long as there is a separate demand for the particular component, then there is a separate product market for that component.[367] Yet complaints have been made that too strict a rule harms IP rights and their owners unnecessarily.[368]

---

[361] See Cases 238/87 *Volvo v Veng* [1988] ECR 6211; C–22/78 *Hugin v Commission* [1979] ECR 1869; T–30/89 *Hilti*; C–53/92P *Hilti*; C–333/94, *Tetra Pak II*; and C–241–42/91P *Magill*.

[362] Case 78/70 *Deutsche Grammophon Gesellschaft mbH v Metro-SB-Großmärketete GmbH &Co* [1971] ECR 487, [1971] CMLR 631, para. 16.

[363] Case T–30/89 *Hilti*, para. 93 and T–51/89 *Tetra Pak II*, para. 23.

[364] Case 238/87 *Volvo v Veng*, paras 8 and 9.

[365] Ibid., and Cases C–241–242/91P *Magill* where the UK and Ireland were identified as the geographic market; this was also the two only countries where TV listings were copyright protected. On the conflict between intellectual property rights and EC law see Sir Hugh Laddie, 'National IP Rights: A moribund Anachronism in a Federal Europe' (2001) 23 *European Intellectual Property Review* 402.

[366] *Lipton Cash Registers/Hugin* [1978] OJ L22/23, [1978] CMLR D19, confirmed in Case 22/78 *Hugin*.

[367] Ibid.; Cases C–241–42/91P *Magill*, paras 52 and 54; T–83/91 *Tetra Pak II*, paras 83–84, and T–30/89 *Hilti*, para 68.

[368] Govaere, *supra* n. 50, at paras 10.05–10.12 and Korah (2006), *supra* n. 14, p. 170.

In both *Hilti* and *Tetra Pak II* the markets were very much defined around the IP right, whereas in *EC Microsoft* the IP right played a much less significant part.[369] The reason for the feared foreclosure in *EC Microsoft* was more due to the fact that Microsoft had market power in the operating system market and network effects in this market, rather than Microsoft's application of its exclusive rights. This could be due to the actual type of the IP right – in *Hilti* and *Tetra Pak II* it was patents, and in *EC Microsoft* it was copyrights. The former is seen as a stronger right than the latter because patents can exclude third parties from making the same product, whereas copyrights can merely exclude others from copying the product, not from producing a similar product.[370] Naturally, this raises questions whether the market definitions in previous case law were merely accidental in following the scope of the IP rights due to the Commission's general policy of a (very) narrow market definition. This definitely seems to be the case as far as *EC Microsoft* is concerned. The cases show that the EC Courts have supported the Commission in its narrow market definitions.

It is important to emphasise the consequences for IP rights holders of such a finding. It almost automatically makes the IP holder dominant in his or her own spare part market, because in principle no one else would be able to produce a similar product, assuming that all spare parts are not substitutable owing to the IP right. Design rights are exceptional in that respect, unless they have been granted a licence by the right holder. This rarity was illustrated in *Volvo v Veng*, which was not a tying case but dealt with the refusal to supply. Volvo held the registered design right on the front wing of the Volvo series 200 cars.[371] Veng imported these front wings, manufactured them without authority from Volvo, and marketed them in the United Kingdom.[372] Volvo sued Veng, which as a defence argued that Volvo was abusing its dominant position by refusing to supply the front wings. The ECJ did not even consider the relevant market but simply stated that the right to refuse was the very subject matter of the IP right, and thus could not 'in itself constitute an abuse of a dominant position'.[373] In other words, the dominance was self-evident. The IP right excluded the possibility of substitutes in this case:[374]

---

[369]   *EC Microsoft* Commission Decision, para. 418.
[370]   Cornish, W.R. and Llewelyn, David, *Intellectual Property: Patents, Copyright, Trade Marks and Allied Rights* (5th edn, Sweet and Maxwell, London, 2003), pp. 6–7.
[371]   Case 238/87 *Volvo v Veng*, para. 3.
[372]   Ibid.
[373]   Ibid., para. 8.
[374]   Anderman, *supra* n. 15, p. 160.

As long as components *performing the same function* were available and they were 'interoperable' with Volvo cars, the existence of the design right would not have the effect of precluding substitutes from the product market. If, however, the function of the component is inseparable from its appearance, as in body panels, and the [IP right] gives an exclusive right to products with that appearance, the existence of the [IP right] can have the effect of narrowing down the product market to a single product market. This in turn sets the scene not only for the finding of dominance on that product market for the rightholder; it could also convert a position of dominance into a monopoly[375] [emphasis added].

The relevant market definition will therefore no longer be a condition for the establishment of dominance; instead it can in certain cases be said to be a tool to ensure dominance.[376] This somewhat harsh claim is confirmed by the Commission itself when it states that it is willing to apply Article 82 to safeguard effective competition or even introduce competition into markets where dependence upon other products can easily occur, such as computer software and telecommunications markets, where there is a need for essential infrastructures or specially designed compatible products.[377] It is important to acknowledge the Commission's policy because it has a great influence on companies' ability to tie and bundle products and/or services together in particular in markets of *de facto* monopolies where a product works as an indispensable input for another product in a related market.[378] The Commission has actively sought to ensure that competition remains in the related markets despite the presence of IP rights.[379]

The question left unanswered is whether a complex product which may contain one or two IP right-protected components will always be at risk of being unbundled by the Commission's narrow market definition policy. How fully integrated must the components be before the Commission accepts them as one product? Judging from the current case law, however meagre, the conclusion is that a tying arrangement involving any form of IP right has so far been subjected to narrow market definitions following the scope of the IP right, thereby making the product identical to the IP right. For this reason a detailed analysis of substitutes does not always take place.[380] Despite criticism of the Commission's and the EC Courts' approach to the tying cases, it must

---

[375] Ibid.
[376] See also Monti (2004), *supra* n. 314, p. 31.
[377] See Case T–70/89 *BBC v Commission* [1991] ECR II–535, para. 35 and Anderman, *supra* n. 15, pp. 160–161.
[378] Ibid.
[379] Ibid.
[380] Rahnasto, Ilkka, *Intellectual Property Rights, External Effects and Antitrust Law, Levering IPRs in the Communications Industry* (Oxford University Press, Oxford, 2003), p. 23.

be emphasised that all the cases are of an extreme nature, as regards both the alleged abuse and the degree of dominance found – ironically due in some respect to the IP right involved. The conclusion is therefore that when the patent or other form of IP right is applied to extend the scope of that right itself by binding unprotected products to protected ones, it will trigger the competition laws. Accepting such a conclusion also means accepting that the boundaries of an IP right create the scope of the product market itself. Such a conclusion can be dangerous because the product is (often) distinct from that which the IP right protects. It will thus automatically mean that there is an abusive tie between an IP right-protected component and a non-protected one, even though there may not be a separate demand for the IP-protected component without the non-protected one – or vice versa.[381]

It is important to note that there can be a distinct market for the sale and licensing of IP rights, which can be distinguished from the product markets where products protected by IP rights belong.[382] The Commission has dealt with this issue by recognising that it is the technology covered by IP rights that is a product, and not the IP right in itself, but has done so only under Article 81 and its block exemptions, not Article 82 cases.[383] IP rights protect intangible property; they can protect products, parts of products, or processes to produce a product. Products, on the other hand, are tangible property. Although many have argued that IP rights should be treated on an equal footing with tangible property and thus should not be exempted from the competition rules,[384] IP rights should not receive less favourable treatment than tangible property.

In adopting an excessively narrow view of when a component constitutes a product, that is by the scope of the IP right, the risk is that certain otherwise integrated products suddenly become two, and tied due to a rigid and unrealistic product test. A market definition tailored around the IP right does not actually take into account consumers' understanding of the product, which is the one that should be recognised in tying cases.[385]

Although case law indicates that a finding of a narrow market definition and thereby a sure finding of dominance is often the case when dealing with

---

[381]   O'Donoghue and Padilla, *supra* n. 9, p. 101.

[382]   Anderman, S. and Kallaugher, J., *Technology Transfer and the New EU Competition Rules, Intellectual Property Licensing after Modernisation* (Oxford University Press, Oxford, 2006), s. 6.54.

[383]   The TTBER, *supra* n. 18; see also Anderman and Kallaugher, *supra* n. 382, s. 6.55.

[384]   Rahnasto, *supra* n. 380, p. 49, and Ritter, C., 'Refusal to Deal and "Essential Facilities": Does Intellectual Property Require Special Deference Compared to Tangible Property?' (2005) 28 *World Competition: Law and Economics Review*.

[385]   See O'Donoghue and Padilla, *supra* n. 9, p. 101.

IP rights, both the Commission and the EC Courts have declared that the mere ownership of an IP right does not amount to dominance. The ECJ did so in *Deutsche Grammophon:* 'a manufacturer of sound recordings who holds a right related to copyright does not occupy a dominant position within the meaning of [Article 82] merely by exercising his exclusive right to distribute the protected articles'.[386] Even though this statement makes it clear that normally an IP right will not encompass a dominant position, it does not preclude the application of Article 82 and the fact that an IP right can occur simultaneously with a dominant position or even a monopoly.[387] Importantly, IP rights are negative rights, that is rights aimed at excluding others, and not positive rights for exploitation of the protected product.[388] Successful exploitation of the product will depend upon consumer demand and competition within the market.[389] The Commission holds much the same opinion in its *Discussion Paper*, noting that forcing a dominant company to license its IP rights, even in return for reasonable royalties, would infringe the essence of the exclusive right.[390] The Commission therefore concludes that the actual refusal is not in itself an abuse, but notes 'the mere fact that the refused input is a licence to a valid [IP right] protected by law is not in itself an objective justification'.[391] This also appears to be the line of reasoning that the Commission took in its *Article 82 Guidance* although the wording of this document is much more vague.[392]

It would be expected that the IP holder would, to a certain degree, be able to behave independently due to the protection granted and thus the uniqueness of the product. This is further emphasised by the fact the IP rights will be used as indicators of dominance, as was the case in both *Hilti* and *Tetra Pak II*.[393]

IP rights raise a distinct problem in relation to abuse as the very core of the IP right is the right to exclude others from making, selling and using the protected product or service.[394] The ECJ has accepted this by incorporating into its case law the right to safeguard IP rights. This was seen in *Volvo v Veng* where the ECJ acknowledged such a right and held that exclusion could not be seen as an abuse of a dominant position.[395] Furthermore, not all IP rights are created by Community law; many, such as patents, are still established under

---

[386] Case 78/70 *Deutsche Grammophon*, para. 16.
[387] Case C–311/84 *Télémarketing*, para. 16, and Anderman, *supra* n. 15, p. 169.
[388] Ibid.
[389] Ibid.
[390] Commission *Discussion Paper*, *supra* n. 14, p. 67.
[391] Ibid., p. 67 n. 141.
[392] *Article 82 Guidance*, *supra* n. 6, pp. 22–24.
[393] Cases T–30/89 *Hilti*, para. 93 and T–51/89 *Tetra Pak II*, para. 23.
[394] Cornish and Llewelyn, *supra* n. 372, p. 6.
[395] Case 238/87 *Volvo v Veng*, para. 8.

national law; hence, the ECJ is not entitled to rule on whether the actual IP right is appropriate for the invention it protects.[396]

However, if an IP holder applies the IP right to expand the width of his core right, for instance by tying an IP right-protected product with a non-IP right-protected product, it will be seen as abusive under Article 82, because the IP right is used to shield not only the protected product but also the tied product. Put like this, it seems in principle fair that the IP holder is restricted from applying such measures, but what if the IP right protected only a part or a component of a product? Currently, it would be an abuse under Article 82, but would it be reasonable? It is here worth noting that the US misuse doctrine has distinguished between the tie of non-staple goods and not staple ones.[397] A similar distinction does not exist under EC competition law. In fact, the ECJ has applied the strict approach vigorously both in tying cases such as *Hilti* and in refusal to supply and licence cases such as *Magill*. The ECJ has therefore been prepared to use Article 82 to restrict the use of IP rights where the rights have been extended beyond their own boundaries and hinder effective competition in a related market.[398]

Some level of foreclosure is likely to occur as a result of the application of IP rights, because exclusion is the very essence of the right. Depending on the type of IP rights, this exclusion right may be stronger or weaker.

A noteworthy point in the tying cases described above is that the IP right as such did not have much influence on the outcome of anti-competitive effects. They seem to have been taken for granted as part of the finding of dominance, but do not seem to have played a part in the assessment of anti-competitive effects. This would suggest that IP rights are not as crucial to whether the tying claim is successful or not. Moreover, it also indicates that IP rights may have been given credit for more market power than they in reality held. Commentators have argued that *EC Microsoft* was in fact not an IP right case.[399] Although they have done so mainly in respect of the refusal to supply section of the case, this argument could be extended to the tying section, as the

---

[396]   See Forrester, Ian S., 'Compulsory licensing in Europe: a rare cure to aberrant national intellectual property rights?', presentation at 'Competition and Intellectual Property Law and Policy in the Knowledge-Based Economy: Comparative Law Topics', Department of Justice/Federal Trade Commission Hearings, Department of Justice, Great Hall, Washington, DC, 22 May 2002, pp. 11–12, Korah, Valentine, 'The Interface Between Intellectual Property and Antitrust: The European Experience' (2002) 69 *Antitrust Law Journal* 801, pp. 9–10 (article downloaded from Lexis-Nexis); and Sir Hugh Laddie, *supra* n. 365, p. 405.

[397]   Chapter 5, Section 2.

[398]   Anderman, *supra* n. 15, p. 188.

[399]   Delrahim, Mekan and Murphy, Frances, comments given at 'The Microsoft Judgment' conference, 25 September 2007, BIICL, London.

copyright held on either product did not play a significant role in the assessment of the tying abuse.

Case law has acknowledged that the exclusive right granted by the patent, copyright or design right and so on is accepted as an objective justification, particularly in relation to refusal to supply or license.[400] However, such a defence can only be applied when the application of the IP right does not go beyond normal exercise or the boundaries of the exclusivity.[401] The conduct of tying thus encounters difficulties when involving IP right-protected products, as tying a protected product (or component) together with a non-protected product (or component) would be seen as going beyond the boundaries of the exclusivity offered by the IP right. Playing the IP right card as a defence is therefore unlikely to persuade either the Commission or the EC Courts to permit the tie.

To conclude, the presence of an IP right seems to play a significant role in the outcome of the relevant market definition and the establishment of dominance, but appears less crucial to the finding of abuse. This could lead to the assumption that IP rights are merely incidental to the outcome of tying cases; however, it should be recalled that the relevant market definition can in effect be the decisive factor in establishing abuse. In other words, there are grounds for concluding that in certain circumstances the tying of IP rights protected products is more likely to be found to infringe Article 82. These certain circumstances could be linked to the type of IP right. It was clear that the copyright in *EC Microsoft* was less influential upon the outcome of the case in comparison to the patents in *Hilti* and *Tetra Pak II*.

5.   LESSONS TO BE LEARNED FROM EC CASE LAW ON TYING

It must be recognised that in comparison to the US there has been limited case law on tying in Europe, and this naturally distorts the picture. Moreover, only a few cases have made it through to the EC Courts, meaning that the current picture of the EC approach to tying is very much influenced by the Commission.

The early cases such as *Napier Brown/British Sugar*, *Hilti* and *Tetra Pak II*, illustrate a strict approach to tying almost amounting to *per se* illegal. Although the Commission and the EC Courts allowed for the hearing of objective justifications for the ties, these were all dismissed for not outweighing the

---

[400]   Case 238/87 *Volvo v Veng*, para. 8.
[401]   Case T–70/89 BBC, paras 53–54.

anti-competitive effects. This was further enforced by the EC Courts' strict interpretation of commercial usage. In particular the ECJ's statement in *Tetra Pak II*[402] made it clear that the EC Courts were unlikely to take a more lenient view of tying. Even though the ECJ may have made the correct decision with regard to Tetra Pak's tie, its statement has made it very difficult for a dominant company to justify a tying arrangement.

While *EC Microsoft* in many respects reinforces earlier cases, the Commission did allow for a more extensive analysis of foreclosure effects and potential pro-competitive effects of the tie, which could suggest the beginning of a more lenient approach to tying at Commission level. This is, for instance, the opinion of Dolmans and Graf, who glorify the EC competition law approach to tying as well-balanced and '[applied] properly, the five-pronged-test will lead to much more refined and economically reasonable results than any per se rule – be it per se illegality or per se legality – could hope to achieve'.[403] They continue by noting that *EC Microsoft* offers a good example of the application of the test.[404] *EC Microsoft* could also be interpreted as creating a distinction between contractual tying and technological integration, allowing for a more lenient approach to technological tying in comparison to contractual tying, or at least, where special market conditions are present, a more thorough assessment of the tying arrangement is called for.

However, the more cynical scholars would argue that *EC Microsoft* was not about developing a softer approach to tying (or technological integration), but rather the Commission's attempt to ensure that its analysis was thorough enough to find Microsoft guilty of infringing Article 82. The CFI ruling sent equally mixed signals as it confirmed that the old fashioned, almost *per se* illegality still applies, but it left the door ajar for technological integration by suggesting the showing of 'superior technology' to defend this particular form of tying. However, the standard of proof for such showing remains unresolved.

EC Microsoft highlights the major difficulties of the current approach to tying arrangements. Firstly, the competition authorities' definition of the relevant market is not sophisticated enough to consider more technologically complex products, and this then affects the consumer demand test, and eventually the assessment of whether the company holds a dominant position in a market. Secondly, the ECJ's interpretation of abuse has made it difficult, if not impossible, for a company to escape Article 82 once it has been found dominant. This difficulty increases with the level of market power. The fact that the concept of competition on the merits is vague and confused, and the threshold for objective justifications is very high, ensures that once a company has been

---

402   Case C–333/94 *Tetra Pak II*, paras 36–37.
403   Dolmans and Graf, *supra* n. 27, pp. 243–244.
404   Ibid., p. 244.

found to possess market power, tying is illegal *per se*. However, on a positive note, all cases so far have revolved around companies holding market shares above 60 per cent in the tying product market, and in both *Hilti* and *Tetra Pak II* the companies held high market shares in the tied product market as well. This leaves a presumption that tying will only be found illegal by the competition authorities if the market power is significant. Such a conclusion is in line with the economic findings of Chapter 2. That said, caution should be exercised as there is only a small sample of case law to justify such a presumption.

Thirdly, the anti-competitive effects analysis and the objective justifications seem to have little influence upon the conclusion as to whether the tying is an abuse or not, as in their current form they do not offer a true picture of the tying arrangement. The anti-competitive effects will always outweigh the objective justifications, owing to the difference in the level of burden and standard of proof. Hence, what seem to be the determinative factors in deciding whether a tying arrangement is abusive or not are whether there are one or two products and whether the company holds a dominant position; once this has been established the tying arrangement is condemned. This is very close to an illegal *per se* approach. Arriving at such a conclusion cannot be a goal to aspire to – it eliminates all legal certainty and indicates that big is bad, in contrast to the wording of Article 82, which holds that it is the abuse of the dominant position that is bad and not the dominance itself. Moreover, it seems to contradict the overall policies of EC competition law. The overall goal of the Lisbon Agenda is making 'Europe . . . "the most *competitive* and dynamic knowledge-based economy in the world . . ."'[405] To put it frankly, how can one expect European companies to compete aggressively with the rest of the world if they are constantly held back by their own competition authority and in fear of fines and remedies because they have grown too big?

The assessment of the remedies indicates that although in most situations the remedies available to the Commission will adequately cease the abusive behaviour and protect or reinforce the competitive process and innovation within the affected markets. However, when it comes to fast-moving industries such as the computer software industry, it becomes a much more complex task to ensure that the remedy both stops the abusive behaviour and contains incentives to innovate within the market. One factor which significantly restricts the possibility of achieving this delicate balance is time. In respect of IP rights, case law paints a depressing picture; firstly, three out of the four best-known tying cases involved IP rights. Secondly, in both *Hilti* and *Tetra Pak II* the IP

---

[405] Communication from the Commission to the Council, the European Parliament and the Economic and Social Policy Committee and the Committee of Regions, 'Social Policy Agenda', COM(2000) 379 final, Brussels, 28 June 2000, p. 2.

rights played a vital role in the market definition and were thus instrumental in the finding of abuse. IP rights played less of a role in *EC Microsoft*, and one could ask whether this was due to the type of IP right.

It is clear that IP rights play less of a role in the abuse assessment and this suggests that IP rights are only incidental to finding tying conduct abusive. That said, with the approach currently taken to tying, where dominance plays a leading role, IP rights are bound strongly to influence the outcome. It can therefore be concluded that when a tying arrangement includes IP rights, it is more likely than not to be found to be an infringement of Article 82.

Overall, this chapter has demonstrated that there is an essential need for re-thinking the approach to tying under EC competition law. The current approach is out of touch with economic thinking as well as market conditions in innovative and fast moving industries and consumer demands.

4. Tying Arrangements under US Antitrust Law

1. INTRODUCTION

US antitrust law has had 120 years to develop. This naturally means that there is an extensive amount of case law and shifting attitudes to tying arrangements taken by the US courts over the decades. However, as this chapter is not a historical account, only cases which have a bearing on the current legal situation will be dealt with in detail.

US antitrust law is built upon two main statutory provisions: sections 1 and 2 of the Sherman Act (hereafter Section 1 and Section 2).[1] The former restricts all contracts or combinations or conspiracies between persons; the latter prohibits any monopolisation or attempt to monopolise. EC competition law's main Articles may appear limited in their wording and lacking in explanation for their application. Sections 1 and 2 offer even fewer hints. Much weight has therefore been put upon the shoulders of the courts to establish judicial standards which fulfil the goals of the antitrust policy.[2] In response, the courts developed the infamous illegal *per se* and the *rule of reason* standards.[3] In particular, the former has very much shaped the current attitude to tying arrangements.

Two more pivotal Acts, the Federal Trade Commission Act[4] and the Clayton Act,[5] play a role in the regulation of tying arrangements. In other

---

[1]   Sherman Act Section 1 and Section 2 (1890), 15 U.S.C. §1 and §2

[2]   Bork, Robert H., *The Antitrust Paradox: A Policy at War with Itself* (The Free Press, New York, 1993), pp. 72, 73–74 and Brief of Prof. Lawrence Lessig as *Amicus Curiae*, in *United States v Microsoft Corp.*, D.C. of Columbia, Civil Action No. 98-1233 (TPJ), 1 February 2000, pp. 6–7.

[3]   See cases such as *United States v Trans Missouri Freight Ass'n*, 166 U.S. 290 (1897); *US v Joint-Traffic Ass'n*, 171 U.S. 505, 19 S.Ct. 25, 43 L.Ed. 259 (1898); *Standard Oil Co. v US*, 221 U.S. 1, 31 S.Ct. 502, 55 L.Ed. 619 (1911); *Chicago Board of Trade v US*, 246 U.S. 231 (1918); and *US v Socony-Vacuum Oil Co.*, 310 U.S. 150, 60 S.Ct. 811, 84 L.Ed. 1129 (1940).

[4]   FTC Act §5(a)(2) as amended (1975).

[5]   15 U.S.C. §§ 12–27, Sections 2, 3, 7, and 8. It is important to note that Clayton Act Section 3 specifically mentions patented products. While there is no refer-

words, there are four provisions which can catch tying arrangements.[6] Most tying cases, or rather the most important tying cases, have generally been assessed under the Sherman Act, mainly because its scope is much broader than, for instance, the Clayton Act section 3.[7] Other cases have been filed under both.[8]

The standard of analysis is the same under both the Sherman Act and the Clayton Act.[9] However, originally it was assumed that a Clayton Act violation occurred if *either* the defendant *enjoyed a monopolistic position* in the market for the tying product *or a substantial volume of commerce* was curtailed. In contrast, for a Section 1 violation both conditions had to be met.[10] Although this distinction has been eliminated, requiring both conditions to be met, Clayton Act section 3 specifically requires the assessor to condemn tie-ins only where substantial lessening of competition could occur. Hence, a strict *per se* approach could never be adopted for cases falling under the Clayton Act section 3. This indicates that if the courts wanted to avoid rule shopping and ensure equal application and outcome regardless of what rule the plaintiff claimed had been violated in respect of tie-ins, a more flexible approach had to be adopted.

Even though case law has mainly made use of it in *rule of reason* analysis, Section 1 in principle condemns contracts and conspiracies only if they are 'in restraint of trade or commerce'. This part of Section 1 matches the requirement under Clayton Act section 3. As courts have relied upon precedents applying different statutory provisions, the development of tying case law under the different statutes has become 'indelibly intertwined'.[11]

---

ence in Clayton Act Section 3 to other forms of IP, there is no doubt, judging from case law, that these will also fall within its scope and can therefore not be used as a defence to tying: Hovenkamp, Herbert, Janis, Mark D., and Lemley, Mark A., *IP and Antitrust, An Analysis of Antitrust Principles Applied to Intellectual Property Law* (Aspen Law & Business, New York, 2007), vol. I, pp. 21–26.

[6]   *United Shoe Machinery Corp. v. United States*, 258 U.S. 451(1922), a case filed under the Clayton Act section 3; *Federal Trade Commission v Gratz*, 253 U.S. 421 (1920), filed under section 5 of the Federal Trade Commission's Act.

[7]   In particular, section 3 does not cover services: Montgomery, W. 'The Presumption of Economic Power for Patented and Copyrighted Products in Tying Arrangements' (1985) 85 *Columbia L. Rev.* 1140, n. 2.

[8]   *International Salt Co. v United States*, 332 U.S. 392, 68 S.Ct. 12, 92 L. Ed. 20 (1947).

[9]   *Moore v Jas. J. Matthews & Co.*, 550 F.2d 1207 (9th Cir. 1977), in which the court stated that the 'standards are "virtually identical"' at 1214 and note Bork, *supra* n. 2, p. 366, although there is still a difference in their jurisdictional scope: Hovenkamp, Janis and Lemley, *supra* n. 5, pp. 21–27.

[10]   Compare *United States v United Shoe Mach. Co. of N.J.*, 247 U.S. 32 (1918) with *United Shoe Machinery* (1922); see also *Times–Picayune Pub. Co. v United States*, 345 U.S. 594, 73 S.Ct. 872, 97 L.Ed. 1277 (1953).

[11]   US Department of Justice, 'Competition and Monopoly: Single-Firm Conduct

The Supreme Court had until recently steadfastly maintained that the test applied to tying should take a strict illegal *per se* approach,[12] however, the so-called *per se* analysis applied to tying takes into account aspects such as market power, harm to commerce, and objective justifications for the behaviour as a defence, in line with the Clayton Act's more lenient language. The market power requirement also aligns the approach with Section 2 and Article 82 of the EC Treaty's assessments of tying,[13] though the few tying cases which do fall under Section 2 will normally be assessed under a *rule of reason* approach rather than the rigorous *per se* illegality under Section 1.[14]

2. THE ANTITRUST ANALYSIS OF TYING ARRANGEMENTS

While Europe has remained almost static in its approach to tying arrangements, the US has made significant developments in its approach to tie-ins.[15] Of major influence have been the economic theories, in particular those of the Chicago School.[16] The focus of this book is on the current situation; early case law will be looked at only briefly. Greater attention will be given to the most recent case law, in particular *Jefferson Parish, Eastman Kodak, US Microsoft II*,[17] *US Microsoft III*[18] and *Illinois Tool*. This section is thus divided into two

---

Under Section 2 of the Sherman Act (2008)' (www.usdoj.gov/atr/public/reports/236681.htm). (*Section 2 Report*), p. 78.

[12]   *Standard Oil Co. of California v US*, 337 U.S. 293, 69 S.Ct. 1051, 93 L.Ed. 1371 (1949); *Northern Pacific Railway Co. et al. v United States*, 356 U.S. 1 (1958); and *Jefferson Parish Hospital District No. 2 v Hyde*, 466 U.S. 2, 104 S.Ct. 1551, 80 L.Ed.2d 2 (1984).

[13]   Ahlborn, Christian, Evans, David S., and Padilla, A. Jorge, 'The Antitrust Economics of Tying: A Farewell to *Per Se* Illegality' [2004] *Antitrust Bulletin*, 287, p. 292.

[14]   *Eastman Kodak Co. v Image Tech. Serv.*, 504 U.S. 451 (1992) and Peritz, R. J. R., 'Competition Policy and Its Implications for Intellectual Property Rights in the United States', chapter 3 in Anderman, Steve (ed.), *The Interface between Intellectual Property Rights and Competition Policy* (Cambridge University Press, Cambridge, 2007), p. 204.

[15]   *Illinois Tool Work Inc. v Independent Ink Inc.*, 547 U.S. 1 (2006), at 1.

[16]   Sullivan, Lawrence A. and Grimes, Warren S., *The Law of Antitrust: An Integrated Handbook* (Hornbook Series, West Group, St Paul, Minn., 2000), p. 204, n. 55; however note Hylton and Salinger and Feldman, who give a different account of the influence of in particular the Chicago School: Hylton, Keith N. and Salinger, Michael, 'Tying Law and Policy: A Decision-theoretic Approach' (2001) 69 *Antitrust Law Journal* 476–479, pp. 469–470 and 476–479 and Feldman, Robin Cooper, 'Defensive Leveraging in Antitrust' (1999) 87 *Georgetown Law Journal* 2079, pp. 2109–2112.

[17]   *US v Microsoft Corp.*, 147 F.3d 935 (D.C. Cir. 1998) (*US Microsoft II*).

[18]   *US v Microsoft Corp.*, 253 F.3d 34 (D.C. Cir. 2001) (*US Microsoft III*).

parts; the first deals with tying cases pre-*Jefferson Parish* (1984) and the second with tying cases post-*Jefferson Parish* (this part includes a case analysis of *Jefferson Parish* itself). The reason for making a distinction in the case law analysis between case law prior to *Jefferson Parish* and *Jefferson Parish* and succeeding cases is that although *Jefferson Parish* in many ways applies the traditional approach to tying and also confirms the necessary requirements, the Supreme Court disagreed on the name of the approach and therefore opened up the possibility of a more liberal view of tying, which has largely been followed in subsequent cases. This has also been confirmed by surrounding events, for instance the abolition of the Nine no-nos[19] and changes to the US Patent Act by the Patent Misuse Reform Act of 1988. The latter clarified that tying of a patented product was no longer illegal unless market power for the patented product was proved.[20] The amendment corresponds to the Supreme Court's approach in antitrust case law, which requires there to be market power before condemning the tying arrangement. As will be seen, the requirement for market power is also compatible with that found under EC competition law. Section 271(d) of the US Patent Act itself will be discussed in Chapter 5.

2.1. Tying – Pre-*Jefferson Parish*

Initially in *US v United Shoe Machinery*, the Supreme Court accepted that the 'best results are obtained' through a tying arrangement.[21] In this case it was

---

[19]   The Nine No-Nos is a list of certain types of licensing practices which were seen by the DoJ as causing unreasonable harm to competition. The list is 'rather formalistic' (Pitofsky, Robert, 'Challenges of the New Economy: Issues at the Intersection of Antitrust and Intellectual Property' (2001) 69 *Antitrust Law Journal* 913–924, p. 918 and shows an excessive concern to safeguard competition and thereby offset the benefits of innovation. The Nine No-Nos were: (1) tying the purchase of unpatented materials as a condition of the licence; (2) requiring the licensee to assign back subsequent patents (grantbacks); (3) restricting the right of the purchaser of the product in the resale of the product; (4) restricting the licensee's ability to deal in products outside the scope of the patent; (5) a licensor's agreement not to grant further licences; (6) mandatory package licences; (7) royalty provisions not reasonably related to the licensee's sales; (8) restrictions on a licensee's use of a product made by a patented process; and (9) minimum resale price provisions for the licensed products. See Wilson, Bruce, B. Deputy Assistant Attorney Gen., Remarks before the Fourth New England Antitrust Conference, Patent and Know-How License Agreements: Field of Use, Territorial, Price and Quantity Restrictions (6 Nov 1970) and Muris, Timothy J., 'Competition and Intellectual Property Policy: The Way Ahead', American Bar Association Antitrust Section Fall Forum, Washington, DC, 15 November 2001 (www.ftc.gov/speeches/muris/intellectual.htm#N_2_).
[20]   35 U.S.C. § 271(d)(5).
[21]   *US v United Shoe Machinery* (1918) at 64.

the tying of different United Shoe machines and leases which restricted the use of third party machines with United Shoe's machines. However, this liberal view was short-lived, and in 1922 in *United Shoe Machinery*[22] the Supreme Court found more or less the same tying arrangement to infringe Clayton Act section 3. The general presumption therefore, especially in the early case law, was that tying arrangements were illegal *per se*, and a strong resistance was shown towards tying agreements.[23] This was the case regardless of whether IP rights were involved or which rule was invoked. In *United Shoe Machinery, IBM Corp.*[24] and *International Salt*, the defendants had been accused of making the lease of a product conditional upon the sale or lease of another. All three claimed that the IP right held in respect of the tying product allowed them to tie. However, the Supreme Court found all three tying arrangements illegal.

IBM Corp. is interesting in two significant respects: firstly, it introduced the leverage theory as the main cause of concern for tying arrangements;[25] and, secondly, it entrenched a high threshold for the objective justification defence.[26] In *Standard Oil Co. of California*, in which the Supreme Court infamously held '[tying] agreements serve hardly any purpose beyond the suppression of competition'.[27] Bork remarks on this statement from the Supreme Court, '[this] remarkable assertion has never been supported either theoretically or empirically, and it has become increasingly evident, outside the world of judicial economics, that if there is one purpose a tie-in does not serve it is the suppression of competition'.[28] The Supreme Court finally discarded the *Standard Oil* dictum in *Illinois Tool* in 2006.[29]

Early case law demonstrated that four conditions must be met before a *per se* violation can be established. These elements seem to have stuck over the years despite the change in attitude towards tying arrangements,[30] albeit that they have altered somewhat as regards which elements have been given more attention. The four conditions are:

---

[22]   *United Shoe Machinery* (1922).

[23]   See *Standard Oil Co. of California*.

[24]   *International Business Machines Corp. v United States*, 298 U.S. 131, 56 S. Ct. 701, 80 L.Ed 1085 (1936).

[25]   Hylton and Salinger, *supra* n. 16, n. 28.

[26]   Ibid.

[27]   *Standard Oil Co. of California* at 305–306; this statement was repeated in later cases such as *United States Steel v Fortner Enterprises*, 394 U.S. 495 (1969) (*Fortner I*) at 503.

[28]   Bork, *supra* n. 2, p. 367.

[29]   *Illinois Tool* at 2.

[30]   Waelbroek, Denis, 'The Compatibility of Tying Agreements with Antitrust Rules: A Comparative Study of American and European Rules' [1987] *Yearbook of European Law* 39, p. 42.

(1)   A tying arrangement,
(2)   Market power in the tying product market,
(3)   A substantial amount of commerce in the tied product market, and
(4)   Lack of justification and defences.

2.1.1.   A tying arrangement

For there to be a tying arrangement, there need to be two products as well as evidence of coercion. In respect of the former it would be natural to assume that two product markets, one for the tying product and one for the tied product, are identified. However, the *per se* rule does not contain a specific condition to identify the tied product market, and no specific standard has been developed.[31]

In early case law the requirement was seen as 'purely linguistic' because no tie exists without two products.[32] Early illegal *per se* case law therefore seems to be focused upon contractual tying of complementary products and not physically or technologically integrated components of products.[33] Only in *Times–Picayune* did the Supreme Court question whether there really were two products; this was whether morning newspaper advertisement was a separate product from evening newspaper advertisement.[34]

In the 1970s in a series of technological integration cases the US courts placed a high standard of proof upon the plaintiff such that, in effect, a safe haven was created for technological integration.[35] In *Response of Carolina*,[36] the Court stated that a plaintiff must show that the integration was 'for the purpose of tying the products, rather than to achieve some technologically beneficial result'.[37] As an explanation for this special treatment of technological integration in comparison to contractual tying, Hylton and Salinger state:

---

[31]   Hovenkamp, Herbert, 'IP ties and Microsoft's rule of reason' [2002] *The Antitrust Bulletin*, 369, p. 406 and Ahlborn, Evans and Padilla (2004), *supra* n. 13, p. 294.

[32]   *US Microsoft III* at 85.

[33]   Ibid., at 90.

[34]   *Times–Picayune* at 613.

[35]   Hylton and Salinger, *supra* n. 16, p. 479, see the *IBM* cases: *Cal. Computer Prod., Inc v IBM*, 613 F. 2d 727 (9th Cir. 1979), *Innovation Data Processing, Inc. v IBM*, 585 F. Supp. 1470, 1476 (D.N.J. 1984), *ILC Peripherals Leasing Corp. v IBM* 448 F. Supp. 228 (N.D. Cal. 1978) , *aff'd per curiam sub nom. Memorex Corp. v IBM*, 636 F. 2d 1188 (9th Cir. 1983) (*ILC Peripherals Leasing I*), *Telex Corp. v IBM*, 367 F. Supp. 258 (N.D. Okla. 1973), *rev'd on other grounds*, 510 F. 2d 894 (10th Cir. 1975).

[36]   *Response of Carolina, Inc. v Leaso Response, Inc.* 537 F. 2d 1307 (5th Cir. 1976).

[37]   Ibid., at 1330.

the early technological integration cases introduced, or perhaps brought into sharp relief, an especially protective approach toward product integration in classical tying doctrine. There were two motivations. One is that an uncertain doctrine that threatens harsh penalties for integrating products could deter innovation, an important competitive force. The other concern is that where the advantage or efficiency is in the product design itself ('integration' versus 'bolting'), courts should be especially reluctant to impose liability.[38]

The concern as to why the loophole exists was reaffirmed by the Court comments in *ILC Peripherals Leasing II:* '[w]here there is a difference of opinion as to the advantages of two alternatives which can both be defended from an engineering standpoint, the court will not allow itself to be enmeshed "in a technical inquiry into the justifiability of product innovations".' [39] The statement demonstrates that the courts are reluctant or find it difficult to assess technologically advanced products, such as software products.[40]

Pre-*Jefferson Parish* case law therefore treated contractual tying and technological integration differently. Contractual tying was clearly illegal *per se*, whereas technological integration received more lenient treatment.

2.1.2. Market power

The market power element is the odd one out as a pre-requisite for *per se* illegality of tying. In early case law, there appears to be some inconsistency as to the level of market power required. In *Times–Picayune* the Supreme Court held that the standard for market power was a 'monopolistic position',[41] whereas in *International Salt* the market power was assumed based on the fact that International Salt's salt processing machine (the tying product) was patented.[42]

In *Northern Pacific Railway*, NPR owned 40 million acres of land in the Northwestern states and territories. Upon selling and leasing the land NPR included a preferential routing clause which obliged the purchaser or lessee to use NPR for transportation of goods produced or manufactured on the lands, unless the purchaser or lessee found cheaper alternatives.[43] Such a get-out clause, which was similarly applied in *International Salt*, puts a question mark on the amount of force the customers were really faced with.[44] Yet, the

---

38　Hylton and Salinger, *supra* n. 16, pp. 480–481.

39　*ILC Peripherals Leasing Corp. v International Business Machines Corp.*, 458 F. Supp. 223, 439 (N.D. Cal. 1978) (*ILC Peripherals Leasing II*).

40　See *US Microsoft II* at 950 and Lessig, *supra* n. 2, p. 1.

41　*Times–Picayune* at 608.

42　Hylton and Salinger, *supra* n. 16, p. 475 and *International Salt* at 395.

43　*Northern Pacific Railway* at 3.

44　Hylton and Salinger, *supra* n. 16, pp. 475–476. It can be argued that such a clause permits the seller indirectly to mind competitors' prices. A similar clause was

Supreme Court chose to ignore the clause,[45] and found that NPR had significant market power[46] despite the 'monopolistic position' prerequisite referred to in *Times–Picayune*, holding instead that 'anything more than sufficient economic power' [47] was adequate to cause anti-competitive harm to the tied product market.[48] It is seriously questionable whether this conclusion was correct because of the type of product the tying product was – land.[49] Writing the dissenting opinion, Justice Harlen noted that there is no uniqueness about land in comparison to the uniqueness found in a patented product (here contrasting the case with *International Salt*), and thus an additional assessment of market shares, product superiority and use of tying clause was warranted before it could be established that NPR had market power.[50]

It is clear from *Northern Pacific Railway* that the sufficient economic power requirement in the tying doctrine is not the same as the market power traditionally required under Section 2 for monopolisation, but in fact a lower threshold.[51] It could even be argued that in both *Northern Pacific Railway* and *International Salt* the pre-requisite of significant market power was more a formality finding than a separate element requiring a thorough analysis.[52] Importantly, it did not need to be based on a specific level of market shares. [53] This was confirmed in *Loew's*,[54] where the Supreme Court held that it was not necessary to establish market power in the Section 2 sense;[55] the economic power could derive from the 'uniqueness' of the tying product, which in this case was copyright protected, and consumer desirability.[56] In fact, Loew held only 8 per cent of the film licensing market.[57]

In comparison, the Supreme Court deemed 40 per cent of the New Orleans newspaper advertising market insufficient to confer market power in

---

found to be abusive in the EC *Hoffmann-La Roche* case: Case 85/76 *Hoffmann-La Roche v Commission* [1979] ECR 461, [1979] 3 CMLR 211, paras. 89–90.

[45]   *Northern Pacific Railway* at 7–8; see also *International Salt* at 397.
[46]   *Northern Pacific Railway* at 7.
[47]   Ibid., at 11.
[48]   Ibid.
[49]   Bork, *supra* n. 2, p. 367.
[50]   *Northern Pacific Railway*, at 18–19.
[51]   Hylton and Salinger, *supra* n. 16, p. 475.
[52]   *Northern Pacific Railway*, at 11 and Peritz (2001), *supra* n. 14, p. 213.
[53]   Ibid., pp. 212–213.
[54]   *United States v Loew's Inc*, 371 U.S. 38, 83 S. Ct. 97, 9 L.Ed.2d 11 (1962).
[55]   Ibid., at 46, n. 4.
[56]   Ibid., at 46.
[57]   See Rahnasto, Ilkka, *Intellectual Property Rights, External Effects and Antitrust Law, Levering IPRs in the Communications Industry* (Oxford University Press, Oxford, 2003), p. 24.

Times–Picayune.[58] Ironically with a narrower market definition of the morning newspaper advertising market being separate from the evening market, Times-Picayune would have been in a dominant position.[59]

The *Northern Pacific Railway*, *International Salt* and *Loew*'s finding of sufficient economic power based on the 'uniqueness' of a company's position should be contrasted with that in *Times–Picayune*, where an approximately 40 per cent market share was insufficient to meet this requirement, and later cases such as *Fortner II* and *Jefferson Parish*, discussed below.

In *Fortner I*, the Supreme Court indicated that it believed tying to be harmful even without market power.[60] It recognised though that there needed to be some element of 'uniqueness' about the company's position to create a harmful effect. Although it was unprepared to require a full-blown market power assessment, it noted that the establishment of an appreciable competitive restraint was seen as sufficient.[61] That said, it can be argued that an appreciable restraint on competition will require some element of market power; thus, even if the Supreme Court was reluctant to admit it, market power would have had to be present. This still leaves open the question of how much market power or sufficient economic power, as well as the question of how much market power an IP right can confer.

The Supreme Court has had a fluctuating attitude to market power, in that initially tying was permissible even between patented and unpatented products.[62] It then changed its view and introduced the patent misuse doctrine in *Motion Picture Patents*[63] and *Morton Salt*[64] without further assessment of whether the patent conferred market powers. Subsequently, it held under antitrust law that tying was illegal *per se* applying a presumption of market power due to a patent.[65] It thereby avoided an in-depth analysis of a market power requirement.[66] Successive cases send mixed messages: *Times–Picayune* indicated that a market share of 40 per cent would not meet the sufficient economic power test. This was altered again in *Northern Pacific Railway* in which the Supreme Court seemed to dismiss parts of the *Times–Picayune* judgment. *Fortner I* confirmed a test of uniqueness which seemed to be a form of economic power below dominance, which could be fulfilled by an IP right.[67]

---

58   *Times–Picayune*, at 611.
59   Ibid., at 628 – see also at 611, 613–614.
60   *Fortner I* at 503.
61   Ibid.
62   *Henry v A.B. Dick Co.*, 224 U.S. 1 (1912).
63   *Motion Picture Patents v Universal Film Mfg. Co.*, 243 U.S. 502 (1917).
64   *Morton Salt Co. v G.S. Suppiger Co.*, 314 U.S. 488, 52 U.S.P.Q. 30 (1942).
65   *International Salt* and confirmed interpretation in *Times–Picayune* at 611.
66   *International Salt* at 395.
67   *Fortner I* at 502–503 and 505 and *Loew's*.

2.1.3.   A substantial amount of commerce in the tied product market

The requirement that a substantial amount of commerce be affected in the tied product market by the tie stems from the wording of both Section 1 ('in restraint of trade or commerce') and Clayton Act section 3 ('may be to substantially lessen competition or tend to create a monopoly in any line of commerce'). In *International Salt*, the Supreme Court made it quite clear that tying is illegal *per se*, even if the effect in the market is limited,[68] and '*it is immaterial that the tendency is a creeping one rather than one that proceeds at full gallop*; nor does the law await arrival at the goal before condemning the direction of the movement'[69] (emphasis added). The statement refers to the wording of Clayton Act section 3, but it also signifies the Supreme Court's fear of leveraging.[70] More importantly, it indicates that the antitrust rules work *ex ante* as well as *ex post*.

The requirement of adverse effect on commerce was confirmed in *Northern Pacific Railway*,[71] which also highlighted the two most feared effects of tying which have been the motivators for the courts, and in particular the Supreme Court, condemning tie-ins. Firstly, there is the risk of leveraging market power from the tying product market to the tied product market, restricting free access to the tied product market.[72] Secondly, tie-ins can have a limiting effect on consumer choice.[73] Although in early case law the Supreme Court identified these two risks as being symptoms of bad tying, it did not apply them as a test for identifying whether a not insubstantial amount of commerce had been affected.

Instead, in *Fortner I* the Supreme Court held that a *not insubstantial amount of interstate commerce* was not necessarily related to the scope of or a share of a market foreclosed, but simply 'whether a total amount of business, substantial enough in terms of dollar-volume so as not to be merely de minimis, is foreclosed to competitors by the tie'.[74] In *Fortner I* this dollar-volume was $190,000;[75] in *Loew* it was $60,000;[76] and in *International Salt* it was $500,000.[77] Giving a specific figure – dollar-volume – when a substantial amount of commerce has been affected does not avoid the question what a substantial amount of commerce is. As can be seen from the case law, the

---

68   *International Salt* at 396.
69   Ibid.
70   Feldman, *supra* n. 16, p. 2081–2082.
71   *Northern Pacific Railway* at 6.
72   Ibid.
73   Ibid.
74   *Fortner I* at 501, quoting *International Salt* at 396.
75   *Fortner I* at 502.
76   *Loew* at 49.
77   *International Salt* at 395.

figure has fluctuated, thus limiting legal certainty for the companies. That said, the idea has some benefits, although a clear threshold must be established to create the legal certainty desired by companies, and it must also relate to the level of commerce in the market concerned.

The question is: is the substantial amount of commerce requirement fulfilled when 30 per cent of the tied product market is foreclosed? Or should this be higher (60 per cent for instance) or lower (15 per cent)? None of the above cases answers this question, although *Fortner I* allowed for a safe haven, and that is when the effect on commerce is 'de minimis'. Case law indicates that some evidence of foreclosure has to be present already. However the Supreme Court clarified in *International Salt* that it would be willing to accept an argument that a foreclosure is likely to materialise in the future if the tying conduct is not brought to an end. This is rather unfortunate from a legal certainty perspective as it can significantly lower the threshold of when a tying arrangement is illegal and risk accepting crying wolf without any clear evidence.

2.1.4.   Justification and defences

IBM Corp. established early on that the threshold for allowing defences to tying is very high. IBM claimed that the purpose of the tying arrangement was to maintain IBM's goodwill and ensure the quality as well as the functioning of the machines could be adversely affected by the use of other cards.[78] The Supreme Court rejected this argument, holding that other companies were equally able to produce matching cards which would not damage the tabulating machine. It pointed out that Clayton Act section 3 does not allow for any exceptions, and hence the Supreme Court was not willing to accept IBM's defence, as there were less restrictive measures which IBM could apply to uphold its goodwill without creating or attempting to create a monopoly.[79] This high threshold has remained the rule in subsequent cases.[80]

Jerrold Electronics[81] reached this high threshold, establishing that tying may be allowed when it is used as a means of entering into a new industry and ensuring quality control.[82] Jerrold Electronics had developed a new antenna system; it did not sell separate parts and conditioned its sale upon also

---

[78]   *IBM Corp.*, at 134 and 138–139.
[79]   Ibid., at 139–140.
[80]   See *International Salt*, at 396–398 and Hylton and Salinger, *supra* n. 16, n. 28.
[81]   *United States v Jerrold Electronics Corp.*, 187 F.Supp. 545 (E.D.Pa.1960), affirmed *Jerrold Electronics Corp. v United States*, 365 U.S. 567, 81 S.Ct. 755, 5 L.Ed.2d. 806 (1961).
[82]   Ibid., at 557.

installing and servicing the system. This was due to its initial experiences in the market, which suggested that if it did not follow through with installation and servicing there was a severe risk of system failure.[83] There was no doubt that Jerrold Electronics held market power, as it had sold 75 per cent of the community antenna systems in the country.[84] Following *Northern Pacific Railway*, the District Court found that the tying was illegal *per se*.[85] However, it set aside the *per se* rule because the market had existed for only a short time and because of the position of Jerrold Electronics in the market. It considered the special circumstances of the market and the fact that Jerrold Electronics was relying on the servicing contracts to get payment for the systems, as few communities could afford to purchase the systems up front.[86] Jerrold Electronics was therefore cleared of tying its sale of cable antenna systems with a service contract while the technology had been in its infancy, but at the time of the trial both the industry and Jerrold Electronics had grown sufficiently no longer to require that kind of protection; hence the Government was permitted an injunction.[87] The case illustrates that there is an exception to the *per se* rule where there are special circumstances surrounding the market and industry involved, in particular performance survival tie-ins, as in *Jerrold Electronics*. It has been suggested that this renders the *per se* rule discretionary, and a court may inquire into the reasonableness of the tie 'whenever it finds that the rule's policy against prolonged and fruitless economic investigation does not apply'.[88] The case is interesting as it clearly allows for a loophole in the otherwise strict *per se* rule – even when the company has market power in the tying product market. There are, however, two points which the case raises but does not answer. Firstly, the tie-in was a performance survival tie, but the question is whether the exemption extends to other types of tie. Secondly, Jerrold Electronics was a relatively small and new company, which raises the questions whether the court applied the exception to aid only small companies or whether this exception is also available to bigger and more established companies in new industries; and whether it could be available for small companies in established markets as well.[89]

In *Times–Picayune* the Supreme Court concluded it was primarily legitimate business plans which had motivated the company to tie,[90] and the tying

---

83 Ibid., at 560.
84 Ibid. at 555.
85 Ibid., at 555 and 559.
86 Ibid., at 555–556.
87 Ibid., at 558, 561 and 571.
88 'Use of Tie-Ins in New Industries, the Notes and Comments' (1961) 70 *Yale Journal of Law* 804, p. 808.
89 Ibid., p. 811.
90 *Times–Picayune*, at 622.

arrangement was therefore justified as normal competitive behaviour. The Supreme Court thereby implied that whilst some forms of tying are illegal *per se*, others could be dealt with under a *rule of reason*.[91] However, it took more than five decades before an Appeals Court applied this distinction.[92]

A clear picture emerges of the Supreme Court's limited patience in listening to and accepting business justifications for tying arrangements, except in very exceptional cases such as *Jerrold Electronics* and *Times–Picayune*. The pre-*Jefferson Parish* case law, although having rejected most business justifications put forward, shows that the Supreme Court is willing to accept them if the very high standard established in *IBM Corp.* is met.

2.1.5. Concluding remarks on pre-*Jefferson Parish* case law

Over the years the Supreme Court altered its view moderately. Although it still held that tying arrangements were illegal *per se*, it treated them differently from price fixing and market division. The Supreme Court has clearly stated that price-fixing and market division are always illegal *per se* without proof of market power or anti-competitive effects, whereas tying arrangements appear to be only illegal *per se* if the company holds sufficient economic power on the tying product market, which will restrict competition in the tied product market.[93] The Supreme Court has also allowed companies to give a reasonable justification for the tying arrangement in *per se* prohibition cases,[94] and courts carved out a safe haven for technological integration.[95] Assuming that tying arrangements have no purpose beyond suppression of competition,[96] it is astounding and rather self-contradictory that the Supreme Court adopted a more flexible *per se* illegality rule involving an element of market power in respect of tying arrangements. It is clear that the Supreme Court clearly intended to apply a strict *per se* rule to tying arrangements, but was never fully confident in its position. It felt compelled to explain its actions by introducing a requirement of market power of which it made half-hearted use.

The early *per se* illegality test is therefore not well balanced and exhibits a certain amount of volatility. The pre-*Jefferson Parish* case law is inconsistent, not only with other *per se* rules, because the purpose of the *per se* rule was to avoid an extensive analysis of the economic situation, but also internally, because it in reality requires the courts to engage in some legal analysis before

---

91   Waelbroeck, *supra* n. 30, p. 44.
92   See the analysis of *US Microsoft III* below.
93   *Northern Pacific Railway* and Waelbroeck, *supra* n. 30, p. 47.
94   *International Salt, IBM Corp.* and Waelbroeck, *supra* n. 30, p. 47.
95   *Response of Carolina* at 1330.
96   See *Standard Oil Co. of California* at 305–306.

the conduct can be held to be illegal.[97] In fact, there is little difference between the tying *per se* approach and a normal *rule of reason* analysis, since the tying approach clearly requires a market power assessment, which has been thought to be the distinctive characteristic of the *rule of reason* analysis.[98]

> The real surviving difference between the *per se* rule of tying arrangements and the general rule of reason case is that once (1) power, (2) requisite foreclosure, (3) separate products, (4) coercion, and (5) a 'not insubstantial' volume of tied product commerce actually foreclosed are proven, *anticompetitive effects will be presumed*[99] [emphasis added].

Yet the next section will show that this gap has been narrowed even further as the *Jefferson Parish* ruling actually contains a requirement to prove anti-competitive effects. The Supreme Court's approach to tying is at the same time both encouraging and disheartening. It is encouraging because it shows willingness to understand that tying arrangements may not be as harmful after all, and therefore they should be treated less harshly than other forms of conduct which fall under *per se* illegality. However, it is disheartening to see that the Supreme Court has not been bold enough to eliminate tying arrangements from this category. This has caused legal uncertainty among companies and a risk of being found guilty of conduct which, with an in-depth assessment, could possibly prove to be harmless to competition and even beneficial to consumers.

2.2. Tying – Post-*Jefferson Parish*

Five cases will be discussed here: *Jefferson Parish*, *Eastman Kodak*, *US Microsoft II*, *US Microsoft III* and *Illinois Tool*. In *Jefferson Parish*, a case from 1984, the Supreme Court finally developed the modified *per se* approach. An acute-care hospital, East Jefferson Hospital, required patients wanting an operation to use an anaesthetist from a company (Roux & Associates) which provided all the hospital's anaesthesiology services. An anaesthetist, Dr. Hyde, who wished to work for the hospital, was denied this opportunity because he was not an employee of the contracted company.[100] The excluded anaesthetist subsequently took action, and in the Federal District Court he claimed that the contract between the hospital and the anaesthesiology company was in violation of Section 1, alleging that there was an illegal

---

97   Feldman, *supra* n 16, p. 2110 who also notes that the *per se* rule used for tying is inconsistent with economic theories on leveraging.

98   Hovenkamp, Janis and Lemley, *supra* n. 5, chap. 21, p. 114.

99   Ibid.

100   *Jefferson Parish* at 2.

tie between the use of operating rooms (the tying product) and the anaesthesiology service (the tied product).[101] The District Court denied relief, holding that the consequences of the contract were minimal. The Court of Appeals reversed the case, finding that there was indeed a tying arrangement, and thus found the contract illegal *per se*.[102] The Supreme Court, on the other hand, unanimously found that the contract was not contrary to Section 1, but disagreed on its reasoning. Four judges favoured a *rule of reason* approach, but five judges maintained the *per se* approach, while acknowledging efficiencies in the tying arrangement. One of the reasons given for retaining the *per se* approach was that '[i]t is far too late in the history of our antitrust jurisprudence to question the proposition that certain tying arrangements pose an unacceptable risk of stifling competition and therefore are unreasonable "per se" '.[103] The Supreme Court also referred to Clayton Act section 3, holding that Congress would not have adopted it if it had not seen tying arrangements as a cause of great concern.[104] Then again, it should be recalled that the Clayton Act was adopted in 1914![105] The Supreme Court seemed also to be aware of this, as it, in fact, stated that tying can be beneficial, both for customers and the seller as a means of competing effectively; and as long as other sellers are free to offer similar packages or the products individually,[106] tying is entirely consistent with the Sherman Act.[107]

The choice of continuing with the *per se* approach towards tying arrangements in *Jefferson Parish* is actually consistent with the Supreme Court's general inconsistent attitude towards tying seen in the old case law, as was concluded above. Feldman argues that the inconsistency continues in post-*Jefferson Parish* cases, because the Supreme Court has been unwilling to move away from 'the old, idiosyncratic per se rule'.[108] He is here referring to *Eastman Kodak*, where the *per se* rule was applied. Kodak had adopted policies to limit the availability to independent service organisations (hereafter ISO) of replacement parts for Kodak's equipment and to make it more difficult for ISO to compete with it in servicing such equipment, after ISO began servicing copying and micrographic equipment manufactured by Kodak. ISO complained, alleging that Kodak had unlawfully tied the sale of service for its

---

[101]   Ibid.
[102]   Ibid.
[103]   Ibid., at 9.
[104]   Ibid., at 10–11.
[105]   The case was under the Sherman Act because it revolved around services, which are not covered by the Clayton Act.
[106]   In economic terms, this is known as mixed bundling.
[107]   *Jefferson Parish*, at 11–12.
[108]   Feldman, *supra* n. 16, p. 2110.

machines to the sale of parts, in violation of Section 1.[109] The case differs from other tying cases, as Kodak held no market power on the actual product market. It held a market share of 23 per cent in the photocopiers market, [110] which is below the 30 per cent market share Jefferson Parish Hospital was found to hold.[111] The question was therefore whether Kodak could have market power in the aftermarket for spare parts (the tying product) and tying the services (the tied product) to the spare parts of the photocopiers. In assessing this, the Supreme Court applied the *per se* approach it used for tying arrangements. Kodak's IP rights may have played some role in the Supreme Court's definition of the relevant markets. Kodak had argued that it only held 'an "inherent" monopoly in parts for its equipment',[112] to which the Supreme Court responded that immunity from antitrust law could not be granted in aftersales service markets, as these markets are considerable and growing industries.[113] Although all manufacturers may possess some elements of market power in their own spare parts markets, the Supreme Court found that there were no reasons why that should permit protection from antitrust law in another, albeit related, market,[114] especially since case law dictates that even where market power has been gained through patents, copyright or business acumen, if it is used to exploit and expand into another market the company becomes liable under antitrust law.[115] *Eastman Kodak* was special because the Court accepted that a single brand could be defined as the relevant market; the case defined an aftermarket for services and spare parts, originating from a product market with interbrand competition.[116] *Eastman Kodak* is aimed at two specific situations rather than a broad application to all aftermarkets, as even the uniqueness of some aftermarkets will not in general amount to near-monopolies and confer market power.[117] The first is where a significant number of potential customers are able to acquire only inadequate or partial information to make an accurate assessment of long-term costs of purchasing/leasing the product.[118] The second is where the seller increases aftermarket

---

[109]   Ibid., at 451.
[110]   Hovenkamp, Janis and Lemley, *supra* n. 5, pp. 21–66.
[111]   *Jefferson Parish* at 7.
[112]   *Eastman Kodak* n. 29.
[113]   Ibid.
[114]   Ibid.
[115]   Ibid.
[116]   Peritz, R., 'Theory and Fact in Antitrust Doctrine: Summary Judgment Standards, single-brand aftermarkets and the clash of microeconomic models' (2000) 45 *Antitrust Bulletin* 887, p. 890.
[117]   Hovenkamp, Janis and Lemley, *supra* n. 5, pp. 21–67 and 21–69 (*Supra fn 5*).
[118]   Ibid., pp. 21–68.

prices after the purchaser has bought into the package and is thus locked in to the product.[119]

The *US Microsoft* saga contains very recent and important assessments of tying as anti-competitive. *US Microsoft I* was settled by consent decree.[120] In *US Microsoft II* the Department of Justice (DoJ) had filed a contempt action claiming that Microsoft had violated the 1994 consent decree by bundling its Internet Explorer web browser to its Windows 95 operating system in its licensing agreements with original equipment manufacturers. The Court of Appeals was compelled to interpret parts of the consent degree, in particular section IV(E)1, which prohibited Microsoft from entering into licensing agreements which required the licensee also to license another product, unless the product was integrated.[121] The Court found that Microsoft had not acted anti-competitively, as Windows 95 and Internet Explorer were not two distinct products, but an integrated product.[122] The case was special as the Court focused its conclusion upon whether Microsoft had violated the consent decree, and not antitrust law in general.[123] That said, it raises some interesting thoughts on how to interpret integrated products as opposed to two products merely bolted together. The Court digressed from the consumer demand test developed in *Jefferson Parish*[124] in order to find that Microsoft had integrated its Internet Explorer browser with Windows.

The issue returned in *US Microsoft III*,[125] where the DoJ yet again challenged Microsoft's integration of its operating system with its web browser.[126] Microsoft was again accused of having tied its web browser, Internet Explorer, to its operating system, Windows, by means of both contractual and technological ties. Microsoft argued that the two components constituted a single integrated product and the relevant market was that of platforms for software applications.[127] The District Court of Columbia had found Microsoft guilty of tying under Section 1 relying on previous cases such as *Jefferson Parish* and

---

[119] Ibid.,
[120] *US v Microsoft Corp.*, 56 F. 3d 1448 (D.C. Cir. 1995), Civil Action 94-1564, 15 July 1994, Final Judgment (www.usdoj.gov/atr/cases/f0000/0047.htm) (*US Microsoft I*).
[121] *US Microsoft II* at 938–939.
[122] Ibid., at 952.
[123] *US Microsoft II* at 946 and Woodrow De Vries, M., 'United States v Microsoft' (1999) 14 *Berkeley Technology Law Journal* 303, n. 17.
[124] *Jefferson Parish* at 21–22.
[125] *US v Microsoft Corp.*, 87 F. Supp. 2d 30 (2000), partly reversed in *US Microsoft III*.
[126] Ibid. and Lessig, *supra* n. 2.
[127] *US v Microsoft Corp.*, 87 F. Supp. 30, 47–50 (D.D.C. 2000), n. 3.

Eastman Kodak.[128] On appeal, the Court of Appeals endorsed a new, more lenient approach to tying arrangements. Based on the conditions laid down in *Jefferson Parish*, the Court of Appeals held that in certain circumstances a modified *per se* approach would not give a sufficient picture of the case, and hence a *rule of reason* should be applied in those circumstances.[129]

US Microsoft III (and *II*), moreover, diverges in two important aspects from previous cases. First, none of the previous cases revolved around physically tied or technologically integrated products.[130] Second, in *US Microsoft III*, Microsoft argued that the 'tie improved the value of the tying product to users and to makers of the complementary goods'.[131] The value improvement defence is similar to the quality defences given in *IBM Corp.*, *International Salt* and *Eastman Kodak*. However, in *US Microsoft III* the Court of Appeals had difficulties assessing whether the quality benefits achieved by the integration of Microsoft's products could be achieved equally satisfactorily by quality standards for different web browser manufacturers.[132] Hence, it was likely that, for once, a company had reached the very high standards the Supreme Court had originally set in 1936 in *IBM Corp.*[133]

Finally, in *Illinois Tool*, the Supreme Court had yet another opportunity to assess tying arrangements involving a patent. Trident and its parent company, Illinois Tool Work, manufactured patented inkjet printheads, patented ink containers and unpatented ink. They sold these products to original equipment manufacturers which were licensed to include them in their printers. The original equipment manufacturers agreed that they and their customers would purchase only ink from Trident and Illinois Tool Work. Independent Ink produced an ink with the same chemical composition as Trident's. Trident sued Independent Ink for infringement, but the claim was dismissed for lack of personal jurisdiction. Independent Ink then sought a judgment against Trident of non-infringement and invalidity of Trident's patent, claiming that Trident had violated Sections 1 and 2.[134] The Supreme Court used the case to clarify its current attitude to tying arrangements and its view on previous case law – in particular, it clarified certain aspects of market power in relation to patents. The case ties the patent misuse doctrine case law together with the antitrust case law neatly, proving firstly that the two are intertwined and,

---

[128]   Ibid.
[129]   *US Microsoft III* at 89; the special circumstances were the physical integration of software within a young industry.
[130]   *US Microsoft III* at 90.
[131]   Ibid.
[132]   Ibid., at 88.
[133]   *IBM Corp.*, at 139–140.
[134]   *Illinois Tool* at 2.

secondly and more importantly, that the Supreme Court recognises that for tying to be illegal evidence of market power must be present, and merely being a patent owner is not sufficient.[135] Yet *Illinois Tool* does nothing to change the other requirements for tying arrangements set down in previous case law.

Similarly to the *per se* illegality approach found in early case law, *Jefferson Parish* held that tying would be unlawful *per se* if the following three conditions were fulfilled. Firstly, there must be a tying arrangement between two separate products; secondly, the seller must have market power; and thirdly, there must be a substantial impact on competition in the tied market.[136] If all three conditions are fulfilled tying is illegal *per se*, unless the company has an objective justification for its behaviour, as established in *IBM Corp.*[137] The Court of Appeals in *US Microsoft III* added an additional condition, that 'the defendant affords consumers no choice but to purchase the tied product from it';[138] including coercion as a separate element. This was merely presumed in previous case law. Interestingly, current case law does not delve into the matter, perhaps on the assumption that there would be anti-competitive effects of the tie only if the other elements, in particular market power, were present and that the tie would be enforceable only if there was market power.[139]

2.2.1. A tying arrangement between two separate products

The *per se* rule does not contain a specific condition to identify the tied product market.[140] In *US Microsoft III*, the Court of Appeals commented upon this missing condition in early case law and noted that the requirement was 'purely linguistic', because no tie exists without two products.[141]

The Supreme Court held in *Jefferson Parish* that 'not every refusal to sell two products separately can be said to restrain competition'.[142] However, the focus in *Jefferson Parish* was no longer on separate products literally or functionally. Instead, the Supreme Court focused on whether there was a separate demand for the tied products, stating that without a sufficient demand for the purchase of the anaesthesiological services from hospital services, there could be no tying arrangement.[143] In order to determine this, the Supreme Court looked at the two products, holding that the anaesthesiology component of the

---

[135] Ibid., at 12, 16–17.
[136] *Jefferson Parish*, at 10–18.
[137] *IBM Corp.* at 139–140.
[138] *US Microsoft III* at 85.
[139] *Jefferson Parish*, at 13–14 and 25.
[140] Hovenkamp (2002), *supra* n. 31, p. 406.
[141] *US Microsoft III* at 85.
[142] Ibid., at 11.
[143] *Jefferson Parish* at 21–22.

hospital package could be offered separately, and was actually billed separately. Moreover, evidence from consumers and surgeons showed that they perceived the anaesthesiology service as separate from other hospital services.[144] The Supreme Court noted that there was indeed a tie between the two separate products, but emphasised that '[there] is nothing inherently anticompetitive about packaged sales'.[145] Only if, as a result of the hospital's market power, patients were forced to purchase Roux's services would the tie be anti-competitive; otherwise the patients were quite free to go to a competing hospital and receive another anaesthesiology service there.[146] Justice O'Connor, concurring with the judgment in *Jefferson Parish*, noted, however, that there could not be said to be two separate services as no patients would consider surgery without anaesthesia, the services were two components of one product, and the hospital would not gain more market power by offering the services in a bundle rather than on their own.[147]

In *Eastman Kodak*, Kodak relied on a similar argument to that of Justice O'Connor, asserting that because there was no separate demand for the spare parts without services, there were not two separate markets.[148] The Supreme Court rejected the claim, concluding that even where tying arrangements are of functionally linked products, where one is of no use without the other, the tie will still be illegal under the antitrust rules.[149] The Supreme Court held that Kodak's product was unique as it was complex business machines – high-volume photocopiers and micrographic equipment – and the spare parts for these, including software, were not compatible with those of competitors.[150] This point is fundamental as it restricted the interchangeability of the spare parts with spare parts from other brands.[151] It went on to note that Kodak did not sell a complete system containing equipment, lifetime service and spare parts for a one-off price.[152] What Kodak did was to provide services after the warranty period either through a contract or on a per-call basis.[153] It held 80 to 95 per cent of the service market for Kodak machines and could therefore impose different prices for equipment, services and spare parts for different customers.[154] There was consequently sufficient opportunity for ISO to

---

[144] Ibid., at 22–24 and n. 36.
[145] Ibid., at 25.
[146] Ibid.
[147] Ibid., at 43.
[148] *Eastman Kodak* at 463.
[149] Ibid., at 463.
[150] Ibid., at 456–457.
[151] Hovenkamp, Janis and Lemley, *supra* n. 5, pp. 21–67.
[152] *Eastman Kodak* at 457.
[153] Ibid.
[154] Ibid.

provide services for Kodak machines, and thus it was logical for the Supreme Court to accept the reasoning that there were in fact three separate markets: one for photocopying machines, one for Kodak's spare parts (the tying product market), and one for the servicing of these machines (the tied product market).[155] The bar for whether the latter two were separate markets, the Supreme Court held, was whether there was a 'sufficient consumer demand so that it is efficient for a firm to provide service separately from parts'.[156] It found that Kodak had sold services and spare parts separately in the past and continued to do so to certain customers.[157] Moreover, the entire high-technology service industry appeared to do the same, and this was thus proof of the existence of a separate service market.[158] The Supreme Court thereby confirmed the *Jefferson Parish* consumer demand test by stating that the consumer demand must be sufficient to make it efficient or economically viable for a company to supply the components separately.[159] If that is indeed the case, the components are separate products. This is an important measurement because it means that even if demand or independent manufacturers are found in the tied market, but these are limited, the tied product may not be seen as separate. The Supreme Court's statement offers the tying company some leeway.

In both *Jefferson Parish* and *Kodak* the Supreme Court relied upon evidence of actual market practices, in relation both to consumer perception and demand, but also from a supply perspective, in particular, the defendant's previous market practice and the practice of other competitors.[160] Yet it should be kept in mind in a developing and innovative market the consumer perspective has its restrictions.[161] The consumer demand test shows a more sophisticated approach to tying by the Supreme Court than in previous case law. There is no doubt that for contractual tying, which most case law so far has revolved around, the test suits its purpose,[162] but the test is not sophisticated enough

---

155 See Ibid. at 458–459.

156 Ibid., at 462.

157 Ibid.

158 Ibid.

159 Ibid. and *Jefferson Parish* at 21–22. The Commission has introduced a similar measurement to the two product requirement in its *Article 82 Guidance*, 'Guidance on the Commission's Enforcement Priorities in Applying Article 82 EC Treaty to Abusive Exclusionary Conduct by Dominant Undertakings', Brussels, 3 December 2008 (http://ec.europa.eu/competition/antitrust/art82/guidance.pdf).

160 *Jefferson Parish* at 22–24 and *Eastman Kodak* at 462.

161 *US Microsoft III* at 92 and Langer, J., *Tying and Bundling as a Leveraging Concern under EC Competition Law* (Kluwer Law International, Alphen aan den Rijn, the Netherlands 2007), p. 87.

162 See Lam, D.K., 'Revisiting the Separate Products Issue' (1999) 108 *Yale Law Journal* 1441, p. 1448.

when it comes to more complex products and technological integration. This statement is justified by the strongly dissenting opinions in *Jefferson Parish* and *Eastman Kodak*.[163] The shortfalls in the separate demand test are more evident in innovative and high technology industries, where companies are less likely to have similar costs structures.[164] This may be the case in two particular situations: firstly, where companies commence tying of two products together, making them work in a novel way, so that, although the product may be known, in combination they have new functionalities. A good example of this is the camera phone or the computer with DVD player. The second situation is where companies bundle products which were previously tied together further downstream.[165] *US Microsoft II* and *III* are both good examples of this and the intricacy and density that are required by the legal assessment to answer the question whether the tying is anti-competitive. These forms of bundling, which could be both efficient and beneficial to consumers, would not be recognised by the separate product demand test developed in *Jefferson Parish*; hence another more refined test is needed for more complex and technologically advanced products and technological integration, as the Court of Appeals recognised in *US Microsoft III*.

As mentioned, *US Microsoft II* revolved around identifying whether Internet Explorer had become an integrated part of Windows. In the consent decree Microsoft was prohibited from entering into a licensing agreement which made the licensing agreement conditional upon 'the licensing of any other Covered Product, Operating System Software product or other product (provided, however, that this provision in and of itself shall not be construed to prohibit Microsoft from developing integrated products)'.[166] The major question posed by the case was what does product integration mean? The Court stated that this proviso should be understood as allowing 'any genuine technological integration', not considering whether any component of the integrated product is sold separately.[167] The Court then went on to define integrated product as 'a product that combines functionalities (which may also be marketed separately and operated together) in a way that offers advantages unavailable if the functionalities are bought separately and combined by the purchaser'.[168] It also recognised that mere tying or bolting

---

163   *Jefferson Parish* at 35 and *Eastman Kodak* at 489.

164   Meese, Alan J., 'Monopoly Bundling in Cyberspace: How Many Products does Microsoft Sell?'[1999] *The Antitrust Bulletin*, 65, pp. 80 and 92.

165   Ibid.

166   *US v Microsoft Corp.*, Civil Action 94-1564, 15 July 1994, Final Judgment, Section IV(E)(i) (www.usdoj.gov/atr/cases/f0000/0047.htm).

167   *US Microsoft II*, at 948.

168   Ibid., at 948.

of products without adding any efficiencies would not constitute integration.[169] The Court clarified that the package offered must be distinct from what the customer could create himself by adding the components together.[170]

In this manner, the Court of Appeals in effect established a two-part test for assessing whether products were integrated or not. However, referring to *ILC Peripherals Leasing II*,[171] it noted that it was not for the courts to assess whether the integration is more sophisticated than its stand-alone products rivals.[172] The first part of the test required the Court to assess when the integration took place – at the design or installation stage;[173] if at the installation stage, then the consumer gains no advantage from the integrated products as opposed to the stand-alone products.[174] The Court found that the integration was 'the creation of the design that knits the two together'.[175] However, just because it is convenient for the manufacturer to produce products together, this does not necessarily make it justifiable to tie the products together in the sale.[176] As regards the second part of the test, which required a finding that the integration actually did something good, the Court held that 'there should be some technological value',[177] but a mere net plus was not sufficient; rather the issue was whether the integration brought *some advantages*.[178] According to the Court, Microsoft met that standard of proof.

Although the Court held that its assessment was in line with previous antitrust rulings,[179] there seems to be inconsistencies with previous cases.[180] The Court's test is not based upon the consumer demand test in *Jefferson Parish*, as the Court stated it was 'not convinced that these indicia necessarily point to separateness'.[181] Instead, it applied the New Product Rationale test

---

[169]   Ibid., at n. 12.
[170]   Ibid., at 949.
[171]   *ILC Peripherals Leasing II* at 439.
[172]   *US Microsoft II*, at 950.
[173]   Lessig, *supra* n. 2, p. 12.
[174]   Ibid.
[175]   *US Microsoft II*, at 952.
[176]   Kühn, Kai-Uwe, Stillman, Robert and Caffarra, Christina, 'Economic Theories of Bundling and their Policy Implications in Abuse Cases: An Assessment in Light of the Microsoft Case' (2005) 1 *European Competition Journal* 85, p. 107, and Hawker, N., 'Consistently Wrong: The Single Product Issue and the Tying Claims Against Microsoft' (1998–99) 35 *California Western Law Review*, p. 14.
[177]   *US Microsoft II*, at 949.
[178]   Ibid., at 950.
[179]   Ibid., at 946 and 950.
[180]   See Hawker, *supra* n. 176, p. 11.
[181]   *US Microsoft II* at 947.

suggested by Areeda *et al.*[182] The Court indirectly rejected the consumer demand test adopted in *Jefferson Parish* on the ground that this would require the Court to embark upon a technological assessment of the integrated products,[183] which the Court found to be beyond the courts' 'institutional competence'.[184] Instead a narrow and differential evaluation should be sought.[185]

The Court's conclusion has been highly criticised,[186] not least by the dissenting Judge Wald, who noted that the majority had created a safe haven for software products by holding that they differed from physical products because they were predisposed separation and integration.[187] Judge Wald highlighted that such avoidance by exemption was not a recommended approach under antitrust law and stated, 'this to me is an argument for closer, rather than more relaxed, scrutiny of Microsoft's claims of integration'.[188] Judge Wald suggested a two-part test to establish whether the integration benefits consumers, which is a test much closer to the *Jefferson Parish* consumer demand test.[189] First, it is necessary to identify whether there are real benefits (or synergies) to consumers from the integration of the two products. Second, it must be ascertained whether there is evidence of a genuine market for the two products provided separately.[190] If the latter is the case, it suggests that there are limited efficiencies to be gained by offering the two products integrated. There is a distinction between product integration and products tied together. A tie of two products can be permitted only if the company does not have market power. In such a case, there is no need to prove integration. A finding of product integration will equally render the tie permissible, as the two products are no longer seen as tied – they are merely components of a product. Therefore, when a company, such as Microsoft, has market power the only permissible form of tying is integration.

The difficulty with Judge Wald's test is the balancing of the evidence; and in particular, identifying how much weight should be given to potential efficiencies in comparison to the consumer demand for the separate products.

---

[182] Areeda, P. Hovenkamp, H. and Elhauge, E., *Antitrust Law, An Analysis of Antitrust principles and their Application* (Little Brown & Company, New York, 1996), at vol. x 1746, pp. 224–229.

[183] *US Microsoft II*, at 949.

[184] Ibid.

[185] Ibid, at 949–950.

[186] See Page, W. H. and Lopatka, J.E., 'The Dubious Search for "Integration" in the *Microsoft* Trial' (1998–99) 31 *Connecticut Law Review* 1251, and Woodrow De Vries, *supra* n. 124; see also Lessig, *supra* n. 2.

[187] *US Microsoft II* at 958.

[188] Ibid.

[189] Note that Page and Lopatka disagree on this point: *supra* n. 186, p. 1268.

[190] *US Microsoft II*, at 958.

Moreover, the test does not truly take into consideration industry innovation and development. In contrast, the majority opinion's test does consider this issue by lending itself to the New Product Rationale.[191] However, the test developed by the Court still is not sophisticated enough, as the standard of proof upon the defendant is too low.[192] Judge Wald is correct in recognising that the courts must take appropriate action and not shy away from the difficult questions. In comparison, the majority's ruling seemed to have been more about avoiding technically difficult questions than solving the dispute as to the interpretation of the consent decree. It resulted in carving out a rather grand safe harbour for technological integration in software markets.[193] The ruling therefore left a question mark over how integration should be interpreted and understood.[194]

There was not long to wait until the issue of integration was back in the courts again in *US Microsoft III*. The facts of the case were given above. The discussion of whether *Jefferson Parish* or *US Microsoft II* should be followed was embraced by Professor Lessig in his brief as Amicus Curiae in *US Microsoft III* and, interestingly, Lessig does not come to a conclusion either. Rather he acknowledges that both approaches can be applied, and offers a third option.[195] His brief indicates that he is in favour of a more hands-on approach than that adopted by the Court of Appeals in *US Microsoft II*.[196] Lessig's third option is a balancing test. It allows two software products to be considered as one, regardless of how they are tied (contractually or technologically), if combined in a new way, unless the bundle is proven to cause anti-competitive harm.[197] The point of the test is to catch those forms of tying which are strategic rather than innovative. Lessig refers to the Same Product Rationale applied by Areeda *et al.*,[198] which argues that if the company is bundling complete substitutes, the two products should be considered as one for the purposes of antitrust law, whereas if the bundle is a partial substitution, then the two products are not integrated or one.[199] The idea is that partial substitutes bundling is more likely to be anti-competitive than complete substitutes and the Same Product Rationale can thus act as a check upon the

---

[191] See above and Areeda, Hovenkamp and Elhauge, *supra* n. 182, at para. 1746, pp. 224–229.

[192] *US Microsoft II*, at 951 and Hawker, *supra* n. 176, p. 17.

[193] Ibid., p. 19.

[194] Different levels of integration have also been suggested/identified, such as deep integration, light integration, close integration: see Lessig, *supra* n. 2, n. 27.

[195] Ibid., p. 44.

[196] Ibid., pp. 35–38.

[197] Ibid., pp. 38–39.

[198] Areeda, Hovenkamp and Elhauge, *supra* n. 182, at para. 1747, p. 232.

[199] Lessig, *supra*, n. 2, p. 40.

New Product Rationale to catch those bundles that are merely strategic and hence anti-competitive.[200] Under this balancing test, Lessig finds that Internet Explorer and Windows are two separate products, as Internet Explorer is only a partial substitute for Windows platform.[201]

The Court of Appeals did not discuss Lessig's test in *US Microsoft III*. Lessig is correct in acknowledging a need for a balancing test. However, it is unclear whether his methods will in reality catch the harmful ties and leave the beneficial ones, as the test does not seem to say anything in respect of components which are not substitutes at all.

The discussion above identifies three important issues. Firstly, technological integration should be distinguished from tying, as the former in effect leaves consumers with just one product, whereas tying still allows a perception of two products. Integration is thus one step further than mere bundling of products. This is significant in relation to the market power assessment and the incentives for companies to bundle and integrate. In other words, as noted in *Jefferson Parish*, two separate product markets must be linked before tying exists.[202] If it is not possible to identify two products then the conduct can only be contested as exclusive dealing, vertical non-price restraint or boycott, all of which interestingly fall under the *rule of reason* approach.[203]

This deduction leads to the second issue. All three assessments of *US Microsoft II*, that is the Court of Appeals' majority opinion, Judge Wald's dissenting opinion and Lessig's Brief, establish that integration requires the identification of something more than mere bolting and the efficiencies derived from such bolting. Yet all three assessments offer different methods to establish whether technological integration has taken place. In particular, Lessig's test takes into consideration the signal that a legal integration test will send to companies, and thus attempts to include a check to ensure that the integration is not merely strategic. Finally, it is clear that a simple screening test and avoiding dealing with technologically advanced products, such as software, as suggested by the Court of Appeals in *US Microsoft II*, is insufficient, especially when a company holds market power. There seems to be no way round it: the courts must engage in a meaningful analysis of the products, if not a full-blown *rule of reason* assessment of the facts, to establish product integration.

Despite its ruling in *US Microsoft II* and Lessig's Brief, the Court of Appeals in *US Microsoft III* decided to revert to the *Jefferson Parish* consumer demand test. The Court of Appeals, while noting that not all tying arrange-

---

200   Ibid., pp. 40–41.
201   Ibid., p. 41.
202   *Jefferson Parish*, at 21.
203   Hovenkamp (2002), *supra* n. 31, p. 400.

ments were bad,[204] held that they hinder direct competition between the products, restrict consumers from choosing the products on their merits and as a result of independent judgement, and essentially lead to foreclosure of competition in the tied product market.[205] The Court found the consumer demand test 'a rough proxy for whether a tying arrangement may, on balance, be welfare-enhancing, and unsuited to *per se* condemnation'.[206]

This line of argument – that the separate demand test works as a proxy not only for foreclosure and market power, but also for welfare-enhancing effects of the tying – is a new development in comparison to *Jefferson Parish*.[207] The Court of Appeals held that it was also necessary to look at the supply side as part of a balancing test, because companies without market power resorted to tying only when 'the cost savings from joint sale *outweigh* the value consumers place on separate choice'[208] (emphasis added). In other words, tying is evidence of 'strong net efficiencies' when undertaken by competitive companies.[209] It is therefore important to assess the common conduct of the market, and if all companies are bundling or tying, then illegality *per se* should be rejected.[210] It is interesting to observe that the Court of Appeals only rejects the illegality *per se* rule, not permitting the tie altogether. This indicates a cautious approach similar to the one seen in previous cases. Although acknowledging the benefits of tying, the Court retains a means to regulate the market if the tying nevertheless appears harmful to competition.

In assessing the facts, the Court of Appeals looked at the other major operating systems manufacturers and found that the common practice was to offer a mixed bundle of the operating system and web browser.[211] However, Microsoft asserted that its Internet Explorer was integrated with its operating system and was therefore different from the other operating systems, as there was no need to remove Internet Explorer if the customer wanted to make use of another web browser.[212] Nonetheless, the Court of Appeals held that Microsoft's claims failed the efficiency requirement under monopoly maintenance, because although the integration of Internet Explorer was, according to Microsoft, innovative and beneficial, it necessitated the constant presence of Internet Explorer.[213] Despite this comment, the Court of Appeals accepted

---

204   *US Microsoft III* at 87.
205   Ibid.
206   Ibid.
207   Ahlborn, Evans and Padilla (2004), *supra* n. 13, p. 299.
208   *US Microsoft III* at 88.
209   Ibid., at 88.
210   Ibid.; see also Areeda, Hovenkamp and Elhauge, *supra* n. 182, at para. 1744d.
211   *US Microsoft III* at 88.
212   Ibid., at 88–89.
213   Ibid., at 89.

Microsoft's argument that the separate product test in *Jefferson Parish* was flawed, and it found that the test could chill innovation by companies refraining from adding new functionalities to their products in fear of being caught under the separate product test, which in turn would harm consumers. [214] The Court of Appeals found the separate product test 'backward-looking and therefore systematically poor proxies for overall efficiency in the presence of new and innovative integration'. [215] It feared that any efficiencies deriving from the integration were unlikely to be sheltered by any of the other tying requirements applied in *Jefferson Parish*, and therefore the separate product test was too simple and likely to harm consumers.[216] Nevertheless, the Court of Appeals concluded that due to the findings of Microsoft's conduct of monopoly maintenance, the product integration of Internet Explorer and Windows was not welfare-enhancing, and hence Microsoft could not escape 'tying liability'.[217]

The ruling by the Court of Appeals in *US Microsoft III* returns to the *Jefferson Parish* consumer demand test, although it acknowledges that the test must be developed to deal with more technologically advanced products. The ruling could also be read as meaning that when dealing with integration of products rather than tying, the conduct should be assessed under a *rule of reason* approach. This does not necessarily mean that the court finds technological integration less restrictive on competition than contractual tying, but rather it requires a more sophisticated balancing analysis than that which the *per se* approach and consumer demand test can offer, which would normally be sufficient for contractual tying. Requesting a *rule of reason* approach to technological integration is contrary to the suggestions made by Ahlborn *et al.* in Chapter 2, who suggested a modified legal *per se* approach to technological tying and *rule of reason* approach to contractual tying.[218] *US Microsoft III* illustrates why Ahlborn *et al.* are probably mistaken – technological integration requires more attention than contractual tying due to the difficulties in balancing the efficiencies and anti-competitive effects as well as the risk of stifling innovation. Therefore it is only natural that it should be assessed in greater detail than contractual tying. The DoJ has adopted a similar approach in its *Section 2 Report*, in which it holds that although it is not willing to see technological integration as *per se legal*, 'great caution should be exercised

---

[214]   Ibid., at 89.
[215]   Ibid.
[216]   Ibid.
[217]   Ibid.
[218]   Chapter 2, Section 5.5 and Ahlborn, Evans and Padilla (2004), *supra* n. 13, pp. 339–340.

before condemning a technological tie', and consequently tying should not be *per se* illegal.[219]

2.2.1.1. Concluding remarks on two separate products Jefferson Parish introduced a simple and effective test: the consumer demand test.[220] This test established with a simple question whether there were one or two products. The question was whether there was a demand for the tied product separate from that of the tying product and the bundle. *Eastman Kodak* refined the question by adding that the demand must be sufficient for it to be efficient for a company to provide the tied product, thereby creating a margin for product innovation towards integration.[221] The test works for basic products and standard contractual tying arrangements, which most cases have so far revolved around.[222]

However once the products become more complex, the test reveals certain flaws, such as not taking rapid innovation and product development in high-technology industries into consideration. This was demonstrated in the *US Microsoft* cases. Previous case law had created a loophole allowing technological integration of products to be treated more leniently than contractual tying arrangements.[223] Indeed, in *Jefferson Parish*[224] Justice O'Connor's opinion implies that US tying doctrine creates an incentive for monopolists to make integrated products rather than selling two complementary products in a bundle.[225] The US case law therefore seems to endorse technological integration rather than condemning it. However, it was not until *US Microsoft II* that the courts engaged more thoroughly in assessing technological integration. The Court of Appeals firstly noted that technological integration required something additional to the mere binding of the two products.[226] However it did not further debate what that would be, but simply assumed that the bundling of Windows and Internet Explorer fulfilled this condition.[227] The

---

[219] *Section 2 Report*, pp. 88–89.
[220] *Jefferson Parish* at 21–22.
[221] *Eastman Kodak*, at 462.
[222] Such as *International Salt, Jefferson Parish, Eastman Kodak*, and *Illinois Tool*.
[223] For instance *Telex Corp., Response of Carolina, Inc., ILC Peripherals Leasing I, ILC Peripherals Leasing II* and O'Donoghue, Robert and Padilla, A. Jorge, *The Law and Economics of Article 82 EC* (Hart Publishing, Oxford and Portland, Oregon, 2006), p. 479.
[224] *Jefferson Parish* at 33–40.
[225] Hylton and Salinger, *supra* n. 16, pp. 478–479.
[226] *US Microsoft II* at 949.
[227] Ibid., at 946, 949–950.

Court left no doubt about its discomfort with having to assess a technologically complex product, and this evidently coloured its ruling.[228] In *US Microsoft III* the same issue was revisited, and here the Court relied much more heavily upon the consumer demand test in *Jefferson Parish,* opting for requiring a *rule of reason* approach to cases of exceptional circumstances.[229] What amounts to exceptional circumstances is unclear. It has been suggested that the software industry in itself was the exceptional circumstances. However, the above analysis indicated that the technological integration was also instrumental to the exceptional circumstances of the case.[230] In particular, the latter interpretation would be more in line with previous case law.[231] In other words, case law has carved out a safe haven for technological integration regardless of the level of market power held, as long as it can be demonstrated that there is no substantial impact on commerce in the tied product market.[232]

2.2.2.   Market power

The next step in the assessment of illegal tying is to identify whether the seller had market power. The market power held by the seller in the tying product market must be sufficient for the seller to exploit his control of that market, and thereby force the buyer into purchasing the package rather than the products separately.[233] The forcing will hinder competition on the merits in the tied product market and coerce the buyers into purchasing something that they would otherwise not have done in a competitive market, and is therefore illegal under antitrust law.[234]

When assessing East Jefferson Hospital's market power, the Supreme Court found that 70 per cent of patients residing within Jefferson Parish actually made use of hospitals other than East Jefferson Hospital.[235] The Supreme Court therefore concluded that there was no illegal tying *per se* because, with only a 30 per cent market share, the hospital would not be able to restrain competition with its tying arrangement.[236] This is not quite a repeat of *Times–Picayune*, but at least it implies a form of safe harbour below a market share threshold of 30 per cent.[237]

---

[228]   Ibid., at 949–950 and n. 13.
[229]   *US Microsoft III* at 89–90.
[230]   A more detailed discussion is found in Section 3 of this Chapter.
[231]   *Telex Corp.* and *ILC Peripherals Leasing II.*
[232]   This conclusion corresponds with the DoJ's attitude in its *Section 2 Report, supra* n. 219, pp. 88–89.
[233]   *Jefferson Parish*, at 12.
[234]   Ibid., at 12–14.
[235]   Ibid., at 7.
[236]   Ibid., at 7, 13–14 and 27.
[237]   This idea of a safe harbour is also in line with the Antitrust Guidelines for the

Eastman Kodak was slightly more controversial because Kodak lacked market power in the primary market, and the case illustrated the consequences of a narrow market definition.[238] *Eastman Kodak* is a legal discussion by the Supreme Court of the leverage theory. According to several economists it is not possible to extract two monopoly profits from leveraging – in this case, by tying the two products together.[239] Kodak argued that due to a lack of market power in the main equipment market it would be unable to extract monopoly prices in the spare parts and services markets as it would affect the sale of Kodak machines on the main equipment market, since consumers would choose products with cheaper life-cycle costs.[240] However, the Supreme Court rejected this line of argument, holding firstly that Kodak's sales tactic to restrict lower priced ISO services had led to an increase in service prices for Kodak customers but there was no evidence of a decrease in equipment sale, as Kodak had argued there would be.[241] Secondly, the Supreme Court found that it was difficult, if not impossible, for customers to gain full information as to the costs of the full life-cycle package[242] offered by Kodak, in order for them to make a realistic assessment of whether Kodak's package was worth pursuing.[243] Therefore, Kodak was incorrect in claiming that it would not be able to extract high prices from the aftermarkets. Thirdly, even if some potential customers were able to make such assessments this would not benefit the customers already locked in to Kodak's products. The lock-in would occur as Kodak's spare parts were not compatible with those of other competitors. Hence, once a Kodak copier had been purchased, the purchaser would have to continue to deal with Kodak in respect of spare parts and allegedly also services. Moreover, if the switching costs were high, the locked-in customers would tolerate a certain degree of price increase in services before changing equipment brand, and the Supreme Court had found evidence that this was in fact happening.[244] The Supreme Court concluded that Kodak had sufficient market power in the aftermarkets to raise prices and exclude competitors.[245]

Licensing of Intellectual Property which suggest a safety zone where four or more companies are present in a market: *Antitrust Guidelines for the Licensing of Intellectual Property*, issued by the US Department of Justice and the Federal Trade Commission, 5 April 1995, Section 4.3, but note the comment made in n. 29.

[238]   *Eastman Kodak* at 454–455.

[239]   See Bowman, Ward S., 'Tying Arrangements and the Leverage Problem' (1957) 67 *Yale Law Journal* 19, and Posner, Richard A., *Antitrust Law: An Economic Perspective* (University of Chicago Press, Chicago, Ill., 1976).

[240]   *Eastman Kodak* at 465–466.

[241]   Ibid., at 472.

[242]   That is machine plus potential spare parts and services.

[243]   *Eastman Kodak* at 473–475.

[244]   Ibid., at 476.

[245]   Ibid., at 477–478.

Kodak had chosen to exploit that market power in a market where customers were locked in, there were high information costs, and a possibility of price discrimination which limited or even eliminated long-term loss to gain instant profits.[246] The evidence put before the Supreme Court of Kodak's market power was sufficient.[247] The Supreme Court has on several occasions highlighted that the judgment was a summary judgment and therefore under a trial with a fuller evidentiary record, Kodak might not have been found guilty of tying.[248] In fact, on remand, the ISOs abandoned the violations of Section 1 claims and only pursued the claims under Section 2.[249] *Eastman Kodak* mainly made its name through the Supreme Court's finding of market power in an aftermarket rather than the actual tying arrangement, as this finding established a burden upon IP holders which had not been seen before, because courts had had a reputation for being more lenient towards IP owners. However, the case can be accepted for one particular reason, and that is that Kodak was accused of tying the service and spare parts as well as refusing to sell its parts to the ISOs. It was thereby attempting to foreclose a market in which there were already independent service providers. Moreover, these ISOs serviced not only Kodak's photocopiers but also those of its competitors. Hence, their conduct was part of commercial usage. Although defining the market so narrowly, the case should still be seen in the context of market conditions on the main product market: taking these into consideration, it seems only reasonable that Kodak should not attempt to foreclose the market.

The element of market power was assessed as part of the claim of monopolisation in *US Microsoft III*. Microsoft was here found to hold a market share of at least 95 per cent of the operating systems market.[250] Moreover, the District Court had relied upon certain barriers to entry as well,[251] in particular, the applications barrier to entry, which created a chicken-and-egg situation in the software market, because consumers prefer an operating system with large numbers of applications written for it, and developers prefer to write for operating systems with large numbers of customers.[252] This creates a snowball effect which ensures that Windows, already dominant, will remain so due to customers' and application writers' preference.[253] The Court of Appeals

---

[246]   Ibid.
[247]   Ibid.
[248]   Ibid., at 486.
[249]   See *Image Technical Services, Inc. v Eastman Kodak Co.* 125 F.3d 1195 (9th Cir. 1997).
[250]   *US Microsoft III* at 51.
[251]   See *United States* v. *Microsoft Corp.*, 97 F. Supp. 2d 59 (D.D.C. 2000).
[252]   *US Microsoft III* at 55.
[253]   Ibid.

accepted the District Court's finding of monopoly power. This was despite Microsoft's argument that the software market is uniquely dynamic, and therefore monopoly power in the software industry could only be proved directly, by assessing the company's conduct to determine whether it holds monopoly power.[254] Microsoft attempted a similar argument in its European counterpart, albeit more well-developed. [255] It was argued in the case that the reason for Microsoft's conduct, that is the restrictive licensing agreements and the tie-in of its web browser, was efforts by Microsoft 'to gain market share in one market (browsers) served to meet the threat to Microsoft's monopoly in another market (operating systems) by keeping rival browsers from gaining the critical mass of users necessary to attract developer attention away from Windows as the platform for software development'.[256] The Court of Appeals concluded that Microsoft had illegally maintained its monopoly.[257] There was therefore no doubt that Microsoft held significant market power on the tying product market – the operating systems market.

In *Illinois Tool*, the plaintiff had not put forward any evidence of market power beyond the argument that the defendant held a patent in the tying product market. The Supreme Court noted, referring to academic literature and *International Salt*, that patents do not necessarily confer market power.[258] It found support for this line of reasoning in the IP Licensing Guidelines and concluded that in all tying cases, irrespective of whether the defendant holds a patent, the plaintiff must prove the defendant has market power in the tying product market.[259]

2.2.3. Substantial impact on competition in the tied market

In *Jefferson Parish*, the Supreme Court held that the law distinguishes between conduct which exploits market power to increase the price of the tying product, where the company enjoys some competitive advantage and which may not necessarily invoke the Sherman Act, and conduct which impedes competition in the tied product market and thereby protects a 'potentially inferior product' from competition on the merits.[260] Therefore, '[there] must be a *substantial potential for impact on competition* [in the tied product market] in order to justify per se condemnation'[261] (emphasis added).

---

254   Ibid., at 56.
255   Case COMP/C-3/37.792 *Microsoft*, [2005] 4 CMLR 965, para. 437.
256   *US Microsoft III* at 60.
257   Ibid. at 80.
258   *Illinois Tool* at 15–16.
259   IP Licensing Guidelines, *supra* n. 237, Section 2.2 and *Illinois Tools* at 16.
260   *Jefferson Parish* at 14.
261   Ibid., at 16.

Jefferson Parish clarifies what was implied in *International Salt*: that a potential risk of foreclosure is sufficient to fulfil the substantial impact on competition requirement, albeit that the impact must still be *substantial*.[262] Nevertheless, it is a test that requires an assessment of the future. It therefore introduces an unforeseeable element that naturally plays havoc with any elements of legal certainty in the current tying assessment.

In comparison, the Court of Appeals in *US Microsoft III* stated that if tying arrangements were assessed under the *rule of reason* the standard of proof would be raised from potential anti-competitive effects to actual effects on the tied product market.[263] Although this step to some degree restores the certainty element, it does so at the cost of placing a significant burden upon the plaintiff, not only in respect of the evidence required, but also because it will require the plaintiff to wait until effects have occurred, meaning that an *ex ante* attack will be made more difficult.[264] When there is no market power and the tying arrangement therefore does not fulfil the *per se* illegality criteria, it is for the plaintiff to prove that there is a substantial impact on competition on a *rule of reason* basis.[265] *Jefferson Parish* moves away from the Supreme Court's previous assessment that dollar-volume commerce has been affected to an assessment of whether competition on the merits has been or is likely to have been affected by the tying arrangement.[266] This is much more in line with the approach applied in the EU.[267]

In *US Microsoft II* and *III*, Lessig identified three phases of the Windows/ Internet Explorer integration: in the first phase, Internet Explorer was offered alongside Windows 95, as a single application programme.[268] In the second phase, which Lessig acknowledges is perhaps not a pure phase, some functionalities of Internet Explorer were organised into modules, which could be called upon by the operating system's programmes and the browser technology would be more easily available for other applications. The third phase of the integration would ensure the existence of the browser technology; without it some non-browser functionalities would not function properly.[269]

Following *Jefferson Parish* the Court of Appeals found three different forms of harm to consumers at the phase three level of integration. First, consumers who did not want Internet Explorer were faced with a design that

---

262   See Section 2.1.3 and *International Salt*, at 396.

263   *US Microsoft III* at 95.

264   The plaintiff must demonstrate barriers to entry and that any anti-competitive effects outweigh the pro-competitive effects: *US Microsoft III* at 95.

265   *Jefferson Parish*, at 17–18, 29.

266   Ibid., at 14.

267   See Chapter 3, Section 2.4.

268   Lessig, *supra* n. 2, pp. 2–4.

269   Ibid.

could cause 'performance degradation, increased risk of incompatibilities, and the introduction of bugs'.[270] Second, for consumers who wanted Internet Explorer, 'Microsoft has unjustifiably jeopardized the stability and security of the operating system'.[271] Finally, a consumer wanting another web browser as a default faced an uphill struggle of 'considerable uncertainty and confusion in the ordinary course of using Windows 98'.[272] Interestingly, Microsoft did not offer a justification for these issues, neither was the Court able to identify a technological justification for the design.[273]

2.2.4.   Objective justification

The Supreme Court observed in *Jefferson Parish* that it was not part of commercial usage to tie the anesthesiological services to the operations; in fact, these were billed separately from the other services the hospital offered.[274] The Supreme Court concluded that when there is no market power present to coerce customers to purchase additional services or products, there are no anti-competitive effects.[275] *Jefferson Parish* indicates that when there is lack of market power this takes away the necessity to justify the tying behaviour.

Kodak gave three business justifications for its behaviour in *Eastman Kodak*. The first was the promotion of interbrand competition by protecting the quality of its services, and thereby preventing customers from using the ISOs.[276] However, the Supreme Court shot the argument to pieces by simply noting that Kodak allowed for self-service, which could hardly be said to be upholding its high quality standards.[277] Equally, Kodak's second and third justifications, reduction of inventory costs and preventing free-riding, were rejected by the Supreme Court. In respect of the latter, Kodak argued that the ISOs were free-riding because they had been unable to enter the market for equipment and parts.[278] The Supreme Court on the other hand found that requiring potential competitors to enter two markets simultaneously created barriers to entry and the ISOs' conduct could not be described as free-riding.[279]

---

[270]   *US Microsoft III*, Microsoft Antitrust Trial Fact Findings, 5 November 1999, Civil Action 98-1232, 1233, para. 173.
[271]   Ibid., para. 174.
[272]   Ibid., para. 171.
[273]   Lessig, *supra* n. 2, p. 5.
[274]   *Jefferson Parish*, at 22.
[275]   Ibid., at 25.
[276]   *Eastman Kodak* at 483.
[277]   Ibid.
[278]   Ibid., at 485.
[279]   Ibid.

The cases indicate that the Supreme Court has not changed the high standards of business justification it adopted in 1936 in *IBM Corp.*[280] It appears that a company is more likely to succeed by claiming it lacks market power than to defend the tying arrangement by giving reasonable business justifications, such as quality assurance. The Supreme Court appears willing to accept, or at least assess, forms of business justification other than just quality; nevertheless no case has so far attempted other reasons. However, Meese notes, '[tying] doctrine's hostility toward affirmative defenses insures that cases of monopoly bundling will almost always stand or fall on the answer to the separate product question'.[281] Although Meese is correct as regards the emphasis which has been placed upon the two product requirement, this is clearly demonstrated in the *US Microsoft* cases; it is not so much hostility towards any defences, but rather the fact that the Supreme Court opted for a very high standard of proof and has stuck with it over the years, albeit having occasionally ignored it when it saw fit.[282]

3.   IP RIGHTS

The presence of IP rights seems only to have played a role in respect of the market power requirement for tying arrangements, but in two manners: first, in respect of the definition of the relevant market, where the relevant product market often has been defined around the scope of the IP protected product allowing only for a few substitutes;[283] second, there has in tying cases been a tendency to presume that IP rights (in particular patents) confer market power, if not monopoly, upon their owners in the product markets affected by the rights.[284]

---

[280]   *IBM Corp.* at 139–140.
[281]   Meese, *supra* n. 164, p. 76.
[282]   See *Jerrold Electronics.*
[283]   *IBM Corp.* (patent), *International Salt* (patent), and *Loew's* (copyright); however, see also the more recent case *Jefferson Parish* at 16; compare with Hovenkamp, Janis and Lemley, *supra* n. 5, pp. 4–45.
[284]   *United States v Paramount Pictures*, 334 U.S. 131 (1948) and *Loew's* at 46, *Jefferson Parish* – and see *MCA Television Limited v Public Interest Corp.*, 171 F.3d 1265, 1278 (11th Cir. 1999) (accepting the *Loew's* presumption that copyright confers power), and Justice Harlen's dissenting opinion in *Northern Pacific Railway.* Compare with *United States v. E.I. du Pont de Nemours & Co.*, 351 U.S. 377 (1956), a monopolisation case under Section 2; Montgomery, *supra* n. 7, p. 1152 and Anthony, Sheila F., 'Antitrust and Intellectual Property Law: From Adversaries to Partners' (2000) 28 *AIPLA Quarterly Journal* 1, p. 4 (www.ftc.gov/speeches/other/aipla.htm). Note that the Supreme Court has never explicitly confirmed this: Hovenkamp, Janis and Lemley, *supra* n. 5, pp. 4–13.

In *International Salt*, the Supreme Court noted, albeit in a footnote, that 'common knowledge' dictates that a patent does not always confer market power; sometimes it covers only a unique part of the product.[285] In comparison, in *Jefferson Parish*, it stated in a passing remark that if a product is protected by a patent, 'it is fair to presume that the inability to buy the product elsewhere gives the seller market power'.[286] The opinion of the concurring judge, O'Connor, J., however, was that a patentee holds no market power as long as there are close substitutes for the patented product.[287] In contrast, the DoJ and the FTC have adopted a clear approach in respect of IP licensing: IP rights will be treated similarly to ordinary property rights under antitrust law, and therefore will not be presumed to confer market power on the rightholder, and the licensing of IP rights will generally be seen as pro-competitive.[288] The Supreme Court has, despite previous inconsistent statements, endorsed this approach in *Illinois Tool* by holding that in all tying cases, including those involving IP rights, the plaintiff must prove that the defendant holds market power in the tying product market.[289] This line of reasoning corresponds with the market power requirement in section 271(d)(5) of the Patent Act 1988.[290]

Importantly and perhaps surprisingly, IP rights do not seem to have played any significant role in the two separate product requirement in case law. This is perhaps due to the tying cases mainly having been of contractual tying, and the literal interpretation applied to the separate product requirement in early case law, or the option of dealing with IP cases under IP law; cases such as *Mercoid* and *Mercoid v Honeywell* demonstrate a more thorough discussion of products and components of products.[291] The IP rights' lack of consequence in the two separate product test certainly stands in sharp contrast to the development under EC competition law.

---

[285] *Northern Pacific Railway* at 10, n. 8.

[286] *Jefferson Parish* at 16.

[287] Ibid. at 37 n. 7 – other examples are *Abbott Laboratories v Brennan*, 952 F.2d 1346, 1354–1355 (Fed. Cir. 1991) where the Court stated 'no presumption of market power from intellectual property right', *cert. denied*, 112 S. Ct. 2993 (1992), in comparison the Court in *Digidyne Corp. v Data General Corp.*, 734 F.2d 1336, 1341–1342 (9th Cir. 1984) held that 'requisite economic power is presumed from copyright', *cert. denied*, 473 U.S. 908 (1985).

[288] US IP Licensing Guidance, *supra* n. 237, Section 2.

[289] *Illinois Tool* at 16 – other recent cases illustrate a similar lenient attitude towards IP rights: see for instance *Independent Service Organisations v Xerox Corp.*, 203 F. 3d 1322 (Fed. Cir. 2000) and *Dell Computer Corp.*, 121 F.T.C. 616 (1996). In both cases the Court refused to discuss the motivation behind the companies' decision to exert their statutory rights.

[290] *Illinois Tool* at 12.

[291] *Mercoid Corp. v Mid-Continent Inv. Co.*, 320 U.S. 661 (1944), and *Mercoid Corp. v Minneapolis-Honeywell Regulator Co.,* 320 U.S. 680 (1944).

4. *PER SE* OR *RULE OF REASON* – THAT IS THE QUESTION!

The dissenting opinions in *Jefferson Parish* and *Eastman Kodak* directly and indirectly called for a change in the assessment of tying towards a *rule of reason* approach, both indicating that the treatment under *per se* illegality was too harsh for tying arrangements.[292] Yet the majority opinion in *Jefferson Parish* had argued that it was far too late in history to alter the approach.[293] Nevertheless, it further developed the four conditions for illegal tying by establishing a consumer demand test which required additional assessment of the facts, which in principle goes beyond what is necessary for a *per se* illegality test.

The *per se* approach as such was not discussed in *US Microsoft II*; however the Court, with its tactic of avoiding a ruling, in principle created a loophole in the *per se* illegality of technological integration, in the case of software products.[294] In *US Microsoft III*, the Court of Appeals considered whether the *per se* approach was in fact inappropriate for the case, and found that this was indeed so. It did so on several grounds. First, it focused on the exceptional circumstance of *US Microsoft III*, namely that the market involved was software.[295] Hovenkamp is of the opinion that the Court of Appeals here suggested that the *per se* approach is market-specific and it based its line of reasoning upon the wording in *Jefferson Parish*, which argued that *certain* tying arrangement pose significant risks of harming competition and are therefore illegal *per se*.[296] However, the view that the Court of Appeals intended to claim that the *per se* approach was market-specific is disputable. Its wording could be read as intending to highlight that there were exceptional circumstances surrounding the case; one of them being the fact that it involved a fast-moving high-technology industry. Another circumstance was that there was limited case law experience in the sort of tying arrangement which was assessed in the case, and the Court of Appeals found that the Supreme Court only after considerable experience classified certain behaviour as *per se* illegal.[297] Accepting Hovenkamp's view would mean that the Court of Appeals

---

[292]   *Jefferson Parish* at 35 and *Eastman Kodak* at 489; Gastle, C and Boughs, S., 'Microsoft III and the Metes and Bounds of Software Design and Technological Tying Doctrine' (2001 6 *Virginia Journal of Law and Technology* 7.

[293]   *Jefferson Parish* at 9.

[294]   *US Microsoft II* at 950.

[295]   *US Microsoft III* at 89–90.

[296]   *Jefferson Parish* at 9, *US Microsoft III* at 89; Hovenkamp *supra* n. 31, pp. 402–403.

[297]   *US Microsoft III* at 90.

would have locked itself into a situation where only software should be treated differently. Such a step would be out of character, given the Court's normal cautious approach. The understanding that the Court of Appeals' comment was merely a recognition of the fact that the case dealt with some novel questions that required a different approach – that is not to say that software ties are less likely to harm competition than any other forms of ties – is more in line with its previous rulings.[298] That said, the Court of Appeals in *US Microsoft III* could have been inspired by its ruling in *US Microsoft II*, which was clearly in favour of avoiding dealing with software markets.[299] The Court of Appeals held that it is now common practice within the software markets to integrate new functionalities into platform software, and therefore 'that wooden application of per se rules' in this case could impede innovation in that market.[300]

Perhaps the Court of Appeals decided upon this route in order to avoid having to assess the product integration and the related technical aspects. This would be in line with *US Microsoft II* and other cases.[301] Alternatively, this statement could mean that the Court wanted a more extensive assessment of the bundle before a clear answer could be provided, recognising that tying can at times create efficiencies, but these need to be clearly identified and weighed against potential anti-competitive effects. The statement makes it clear that the Court found the *per se* rule too intrusive for the software industry because it was a common practice within the industry technically to integrate products and a *per se* ruling could risk stifling innovation and, hence, indirectly, competition. Thus the conclusion must be that if there are other similar industries adopting the same form of practice, these could be equally harmed by the *per se* rule. This conclusion corresponds with the approach taken in *Jerrold Electronics*, namely, that although the tying was *per se* illegal there were special circumstances which made the tying necessary for Jerrold Electronics to survive in the market in the infancy of this market.[302]

The other grounds for finding the *per se* approach inappropriate were, firstly, that the tie under investigation was nothing like tying arrangement in previous cases. Early case law had focused on patented products conditioning sale or leases upon the purchase or lease of an unpatented product, and not products which were physically and technologically integrated.[303] Moreover, the defences in early cases were that of quality, which could be achieved by

---

[298] Hovenkamp *supra* n. 31, p. 375.
[299] *US Microsoft II* at 950, n. 13 and Hawker, *supra* n. 176, p. 19.
[300] *US Microsoft III* at 95.
[301] *US Microsoft II* at 950, *Response of Carolina, ILC Peripherals Leasing II* at 429.
[302] *Jerrold Electronics* at 555–556.
[303] *US Microsoft III* at 90.

lesser means than tying, whereas in this case Microsoft argued that the tie improved the value of the tying product for both producers of complementary products and users.[304]

Third, the Court of Appeals accepted that the separate product test was not sophisticated enough to deal with complex and innovative products, as the test does not assess whether other companies use the same conduct and 'because of the pervasively innovative character of platform software markets, tying in such markets may produce efficiencies that courts have not previously encountered and thus the Supreme Court had not factored into the per se rule as originally conceived'.[305] The statement confirms that the Court of Appeals wishes for a more extensive test than the one provided for under the *per se* rule and does not find the courts ill-suited to deal with the technicalities of technologically advanced products.

The Court of Appeals therefore concluded that the tying aspect of the case should be remanded and assessed under a *rule of reason*.[306] However, this never happened. The tying element of the case was dropped by the DoJ when the case returned to the District Court, in an attempt, as the DoJ, argued to be more time efficient.[307] It nevertheless implies that the DoJ doubted the possibility that Microsoft would be found guilty even under a *rule of reason* analysis. The reason may be the previous dismissive ruling in *US Microsoft II*.

US Microsoft III is special because it is the first case which rejects a *per se* illegality instead of a *rule of reason* assessment, but emphasises that it is the two product/integration question which will require a more thorough analysis.[308] It also acknowledges to a much greater extent than previously seen that some tie-ins can be beneficial, and at least cannot be presumed harmful even under a modified *per se* approach. Some have read the requirement of a *rule of reason* approach as the first step on the road to full legalisation of tying arrangements.[309] Although such a reading is an interesting prediction, it should be borne in mind that *US Microsoft III* dealt with technological integration rather than contractual tying. This fact played a role in the outcome, despite the Court relying on contractual tying case law, and will undoubtedly play a role in future cases.

---

[304]   Ibid., at 90.

[305]   Ibid., at 93.

[306]   Ibid., at 84 and 95.

[307]   Press release, 'Justice Department informs Microsoft of Plans for Further Proceedings in the District Court' of 6 September 2001.

[308]   *US Microsoft III* at 92–93.

[309]   Speech by Jean-François Bellis given at 'The Microsoft Judgment' conference, 25 September 2007, British Institute of International and Comparative Law, London.

In *Illinois Tool*, the Supreme Court dealt with a contractual tying case where the facts were less controversial than in the *US Microsoft* cases. The Supreme Court delved only into the issue of market power. The Supreme Court acknowledged that '[many] tying arrangements, even those involving patents and requirements ties, are fully consistent with a free, competitive market'.[310] The Supreme Court confirmed that proof of market power is essential in all tying cases, thus indicating that tying arrangements are no longer strictly illegal *per se* even in contractual tying cases. It made a distinction between previous cases holding that tying arrangements should be assessed under the conditions developed in *Jefferson Parish* and *Fortner II* rather than the *Morton Salt* and *Loew's* strict illegal *per se* rule.[311] It therefore did not say that the *per se* rule no longer applied, but that the *Jefferson Parish* moderate *per se* rule was to apply, which requires certain conditions to be present, including the evidence of market power, before a tying arrangement can be said to be illegal (*per se*).[312] Reading more into the case would be unwise and rash, although the Supreme Court has here opened the door to a much more lenient approach towards tying arrangements, and in particular those involving IP rights.

5. LESSONS TO BE LEARNED FROM US TYING CASE LAW

5.1. Introduction

Case law development shows that the US courts' attitude towards tying arrangements has changed over the years to a more flexible approach.[313] However, following *Jefferson Parish*, tying is still dealt with under an illegal *per se* approach which reflects neither the legal assessment applied to tie-ins nor current economic thinking.[314] Yet the approach is more lenient than an ordinary *per se* approach, firstly in the requirement of a market power appraisal, secondly in the assessment of whether there are two products, and finally in the need to show that a substantial amount of commerce on the tied product market has been affected. However, the approach has been severely criticised over the years by numerous scholars, all calling for tie-ins to be at

---

310   *Illinois Tool* at 16.
311   Ibid., at 13.
312   Ibid.
313   Ibid., at 1.
314   See in general Feldman, *supra* n. 16, and Hylton and Salinger *supra* n. 16, and Chapter 2, Section 7.

least assessed under a *rule of reason*.[315] Hovenkamp (2002) is right when he observes that the *per se* rule applied to tying is too aggressive and encourages the condemnation of tying despite the potential benefits and without any note-worthy assessment of anti-competitive effects.[316] Criticism of the rule is therefore well placed.[317]

The single most damaging element of tying arrangements seems to be the stringent requirement of applying either a *rule of reason* or a *per se* approach. Both seem to have their shortcomings since a *per se* rule will not properly reflect the efficiencies of the tie and a *rule of reason* approach can in principle leave out the separate product requirement,[318] an essential element of tying and the *per se* approach.

5.2. Two Separate Products

The question whether there are two separate products has moved from a purely linguistic test to a consumer demand test for contractual tying arrangements, assessing whether there existed a demand separate from the bundle and sufficiently great for a company to offer the tied product alone. The final development was in *US Microsoft III*,[319] which recognised that this test also has its discrepancies and limitations. As the tying issue was never pursued any further in *US Microsoft III*, it was therefore never clarified whether Microsoft was guilty of tying under a *rule of reason*, and, perhaps more importantly, there was no clarification of how the courts were to tackle the shortcomings of the consumer demand test in relation to technologically integrated products, besides providing a fuller assessment of the pro- and anti-competitive effects under a *rule of reason* approach. *US Microsoft II*, on the other hand, illustrates the fear that many courts probably have of dealing with technologically advanced products and giving effective rulings; nevertheless it introduced a requirement for assessing technologically integrated products. It asked whether the integration was an improvement on the mere tying, which consumers could not achieve by bolting the products themselves, but it left the question of the standard of proof unanswered. It also made clear, together with *US Microsoft III*, that there are at least two distinctions between contractual

---

[315]   See Bork, *supra* n. 2; Feldman, *supra* n. 16; Hylton and Salinger, *supra* n. 16; Meese, *supra* n. 164; and Hovenkamp (2002) *supra* n. 31.

[316]   Hovenkamp (2002) *supra* n. 31, p. 373.

[317]   Ibid.

[318]   Ibid., p. 409. Under Section 2 a unilateral bundling complaint does not require a finding of two separate products: see for instance *LePage's Inc v 3M*, 324 F.3d 141 (3rd Cir. 2003) and *SmithKline Corp. v Eli Lilly & Co.*, 575 F.2d 1056 (3d Cir. 1978).

[319]   *US Microsoft III* at 94.

tying and technological integration. First, the technological integration should, according to the Court of Appeals, be treated more leniently, depending on market conditions it could be assessed under a *rule of reason* approach. Second, the tests applied in respect of the two forms of tying differ. For both contractual tying and technological integration the *Jefferson Parish/Eastman Kodak* test applies. However, *US Microsoft II* additionally requires proof that the technological integration is more advanced than the two products simply working together. Although in the *Section 2 Report*, the DoJ does not suggest a new and more advanced test to assess the products in questions, it does call for a *rule of reason* approach to technological integrated products to ensure that innovation is not halted by antitrust enforcement.[320]

Considering the fact that innovation has never moved faster, it may not be long before another case takes off where the *US Microsoft* cases left off to confirm the assessment of technological integration. The lesson to be learned from this is that a more advanced test is needed fairly to assess technologically advanced and complex products consisting of several components for each of which there may be a consumer demand, such as an operating system and a web browser.

5.3. Market Power

The Supreme Court has clarified firstly that market power is an essential element of a tying analysis even when assessed under *per se* illegality, as it is needed to coerce customers into purchasing the tied product.[321] Secondly, the Supreme Court has recognised that only when there is sufficient economic power is a tying arrangement likely to be harmful to competition. This is also consistent with section 271(d) of the Patents Act, which condemns tying as misuse only if the patent owner has 'significant market power'. *Jefferson Parish* made it clear that 30 per cent market share and below is not sufficient.[322]

US Microsoft III, however, takes an entirely different approach. Despite finding monopoly power,[323] the Court of Appeals did not find that this in itself was enough for it to adopt a *per se* illegality approach.[324] Instead what counted was the understanding of the product, although, it held that the monopoly power and a finding of monopoly maintenance caused the product integration not to be welfare-enhancing.[325]

---

[320] *Section 2 Report, supra* n. 219, pp. 88–89.
[321] *Jefferson Parish* at 27 and *Illinois Tool* at 16.
[322] *Jefferson Parish* at 7, 13–14 and 27.
[323] *US Microsoft III*, at 85 and 51.
[324] Ibid., at 84.
[325] Ibid., at 89.

The case law is unclear as to what exactly sufficient market power is; it seems to be less than what is required of market power under monopolisation in Section 2. That said, even when the market share is found to be as high as monopolisation under Section 2, the Court of Appeals has ruled that in certain cases a *rule of reason* approach would still be more appropriate than the *per se* illegality approach.[326] In other words, the market power requirement plays second fiddle to the separate product condition.

5.4. Substantial Amount of Commerce Affected in the Tied Product Market

The question whether a substantial amount of commerce has been affected in the tied product market has also developed over the years. The Supreme Court has struggled with numbers and figures, and finally turned to a more liberal test in *Jefferson Parish*.[327] Although a *substantial* impact is required, this need not be present at the time of the lawsuit; the mere likelihood seems sufficient.[328] For this form of test to be accepted an advanced economic assessment of the tying arrangement is clearly called for, as this requirement asks the courts to foresee the future. From a plaintiff's perspective, it means that it does not have to be pushed out of business before it can initiate a lawsuit.

5.5. Objective Justification

The Supreme Court has been extremely consistent in its approach to objective justifications, adopting a very high standard in *IBM Corp.*[329] which it has stuck to ever since. The standard is that if there are less strict measures to uphold quality or other business justifications, then tying should not be applied. Whether the standard had been reached in *US Microsoft III* remains unanswered. The fact that the Court of Appeals did not find Microsoft's integration 'welfare enhancing' would indicate that the standard was not reached.[330] On the other hand, the Court never clearly stated that the benefits generated from the tie could be achieved by less harmful means than the tie, leaving the door open on this matter.[331]

---

326   Ibid., at 89–95.
327   See *International Salt* at 395, *Loew's* at 49, *Fortner I* at 502, and *Jefferson Parish* at 14.
328   Note, however, the Court of Appeals' comment in *US Microsoft III* at 95 – when moving to a *rule of reason* assessment actual effect must be demonstrated.
329   *IBM Corp.*, at 139–140.
330   *US Microsoft III* at 89.
331   Ibid., at 88.

Jefferson Parish also introduces another aspect of the requirement, which was elaborated on in *US Microsoft III*, and that is whether other non-dominant companies in the market are engaged in the same conduct.[332] If that is the case then there are likely to be efficiencies to be gained from the tying arrangement for consumers as well. The lesson to be learned from this is that tying arrangements should be seen in light of the current (and perhaps future) market conditions. This line of reasoning matches the Court's own approach in *Jerrold Electronics*. Moreover, the very high standard adopted naturally makes it extremely difficult for companies to justify their behaviour. This high threshold seems to belong to an era when tying arrangements were seen as illegal *per se*, as opposed to the current modified *per se* approach developed in *Jefferson Parish* and *US Microsoft III*, which recognises that not all tie-ins are harmful.

5.6.   IP Rights

The US courts have not been unwilling to apply antitrust rules to IP rights, although they have been inconsistent in their application, in particular as to whether an IP right confers market power. What has not been clarified is how much market power IP rights must confer to make the conduct of tying harmful. *International Salt*[333] and *Loew's*[334] seemed to suggest that, perhaps when it comes to IP rights, the market power requirement changes to an assessment of the scope of the IP right. If the tying goes beyond the scope of the patent, that is tying unprotected products to the protected product, the tying arrangement will be deemed illegal. *Illinois Tool* partially answered these questions by establishing once and for all that an IP right in itself is not sufficient to confer market power; additional evidence of market power must be provided.[335] The additional evidence was not discussed, but one can assume that a traditional market power assessment applied in ordinary product cases should be equally applied to IP rights cases.[336]

Interestingly, there has so far been no case which in respect of the two separate products requirement was required an in-depth analysis of IP rights and the IP protected product(s), and therefore IP rights have not influenced tying case law on this matter as yet. However, with increased innovation in high-technology markets where IP protection of products is to be expected, cases on this matter should be anticipated. *US Microsoft III* provides only a partial answer to this question.

---

[332]   Ibid., at 93.
[333]   *International Salt* at 395.
[334]   *Loew's* at 46.
[335]   *Illinois Tool* at 12, 16–17.
[336]   This would correspond with the IP Licensing Guidelines, *supra* n. 237, Section 2.

5.7. Conclusion

In the US, *Jefferson Parish* made it clear that a 30 per cent market share was insufficient to establish market power; hence in this respect the two jurisdictions are in concurrence.[337] However, other US cases have been less clear in determining what is sufficient market power; early case law has relatively easily established market power, whereas *US Microsoft III* sets a new trend: despite finding significant market power, it did not give an automatic entitlement to apply the illegal *per se* approach.[338] Together with *Illinios Tool* and Section 271(d)(5) of the Patent Act, *US Microsoft III* indicates that the market power requirement is no more than one of the conditions for establishing tying as an antitrust violation, and what is important is the two product requirement and whether there are the anti-competitive effects which derive from the exploitation of the market power.

The Supreme Court has now acknowledged that tying arrangements are not all anti-competitive and are often part of common business strategies.[339] That said, on the surface, tying arrangements still appear to be illegal *per se* because the Supreme Court has said very little to justify any other interpretation.

Interestingly, case law has created a distinction between contractual tying and technological integration; whether this distinction has been created because of the courts' reluctance to delve into technical assessments of products and tying arrangements or a greater thought for the protection of innovation remains a little hazy. Yet it is clear that the courts have recognised that technological integration requires different, more cautious treatment from that of contractual tying to ensure that innovation is not stifled.[340] To that end the courts have looked to the market conditions specific for the technological integration and suggested the application of a *rule of reason* approach to weigh up the pro- and anti-competitive effects of the tie. Although we are a long way from a full-blown economic assessment of every tying arrangement, it can tentatively be suggested that we have officially moved away from a strictly illegal *per se* approach to tying arrangements and are currently looking at a truncated *rule of reason* approach to tying, which is much more in the spirit of the economic findings in Chapter 2.[341] This is not least due to the *Section 2 Report* from the DoJ, which clearly states that the 'rule of per se illegality for

---

[337]   *Jefferson Parish* at 7, 13–14 and 27.
[338]   *US Microsoft III* at 84.
[339]   *Jefferson Parish* at 16.
[340]   This is also now the official line the DoJ has taken: *Section 2 Report, supra* n. 219, pp. 88–89.
[341]   See *US Microsoft III* at 95.

tying is misguided ...'.[342] The US approach is clearly more flexible and courts are more willing to adapt their approach to tying to the specific situations of the individual case in comparison to the EC competition law approach, thereby appearing more tuned in to market development and product innovation. In EC competition law's defence are the limited number of cases and its comparatively young age.

---

[342] *Section 2 Report, supra* n. 219, p. 90.

5. Intellectual Property Law and Tying – An Alternative Approach

1. INTRODUCTION

This chapter seeks to establish how the laws of IP have pursued tying, and whether the approaches can be drawn upon to develop a new alternative approach which is more flexible towards tying under competition law. Tying arrangements occur in most high technology industries. and often these are structured around various forms of IP rights. For instance, the European Commission has made four negative decisions in respect of tying – three of these involved IP rights.[1] Moreover, the US Supreme Court has dealt with a vast number of tying cases under US IP law and in particular the patent misuse doctrine.[2] The chapter will demonstrate that under US IP law a more flexible approach to tying has been applied, and one which is clearly moved away from the *per se* illegality approach currently applied under both US and EC competition law. IP law therefore plays an essential role in reaching a first best solution to tying.

Under EC law, the only way to pursue tying is from a competition law perspective. There is one simple explanation for this, and that is that not all IP rights are EC rights. Some IP rights have been harmonised within the EU, in particular the protection granted to computer software and databases, which deals with interface issues;[3] however, others, such as patents, are still granted

---

[1]   Case IV/30.787 *Eurofix-Bauco/Hilti* [1988] OJ L65/19 (*Hilti* Commission Decision), *Elopak Italia/Tetra Pak* [1991] OJ L72/1, [1992] 4 CMLR 551 (*Tetra Pak II* Commission Decision), and case COMP/ C-3/37.792 *Microsoft*, [2005] 4 CMLR 965.

[2]   *Motion Picture Patents v Universal Film Mfg. Co.*, 243 U.S. 502 (1917); *Carbice Corp. of America v American Patents Development Corp.*, 283 U.S. 27 supplemental op.; *Carbice Corp. of Am. v American Patents Dev. Corp.*, 283 U.S. 420 (1931); *Leitch Manufacturing Co. v Barber Co.*, 302 U.S. 458 (1938); *Morton Salt Co. v G.S. Suppiger Co.*, 314 U.S. 488 (1942); *Mercoid Corp. v Mid-Continent Inv. Co.*, 320 U.S. 661 (1944); and *Mercoid Corp. v Minneapolis-Honeywell Regulator Co.*, 320 U.S. 680 (1944).

[3]   For instance, EU Computer Software Directive 91/250/EEC, [1991] OJ L122/42 and Directive 96/9/EEC [1996] OJ L2/14 on the legal protection of databases.

and enforced nationally and therefore not under the full jurisdiction of the European Union.

If tying is to be pursued via IP law, this can currently be done only through national law, and then only if this does not conflict with EC law, as the latter enjoys supremacy.[4] Therefore, in Europe on the Community level whenever there is a conflict between the two legal spheres competition law will always prevail over IP law. Even when IP laws become EU-wide rights, the same will hold true. For example, when the Technology Transfer Block Exemption of 1996[5] was introduced, it forced the United Kingdom (UK) to repeal section 44 of the UK Patent Act in 2000. The section prohibited tie-ins and tie-outs.[6] The only means of regulating tying conduct in the UK now is either via the EC competition rules or the national equivalent, the Competition Act 1998.

In comparison in the US, IP rights are covered by the US Constitution and are thus granted on a federal level; in fact copyright and patents share the same constitutional origin: '[t]o promote the Progress of Science and the useful Arts, by securing for limited Times, to Authors and Inventors, the exclusive Right to their respective Writings and Discoveries'.[7] Although the IP laws historically condemned tying, tying has actually been seen as a conventional part of a patent licensing agreement.[8] IP law is therefore an interesting contrast to competition law.

Within IP law itself there are restrictions not only limiting which inventions can obtain protection but also setting an internal balance between competition law and IP rights.[9] The US patent misuse doctrine clearly is one such restriction

---

[4]   Case 26/62 *N.V. Algemene Transport en Expeditie Onderneming Van Gend En Loos v Nederlandse Administratie der Belastingen* [1963] ECR 1 and Case 6/64 *Flaminio Costa v ENEL* [1964] ECR 585, 593.

[5]   Commission Regulation (EC) No 240/96 of 31 January 1996 on the application of Article 85(3) of the Treaty to certain categories of technology transfer agreements [1996] OJ L031/2; the regulation has now been replaced by Commission Regulation (EC) No 772/2004 of 27 April 2004 on the application of Article 81(3) of the Treaty to categories of technology transfer agreements (the TTBER) [2004] OJ L123/11.

[6]   Section 44 (a) and (b).

[7]   U.S. Constitution, Art. 1, §8, cl. 8.

[8]   Cornish, W.R. and Llewelyn, David, *Intellectual Property: Patents, Copyright, Trade Marks and Allied Rights* (5th end, Sweet and Maxwell, London, 2003), p. 278.

[9]   IP rights are usually temporary, exclusive rights, artificially created and granted protection by law, and limited to a certain territory: Schechter, R. and Thomas, J.R. *Intellectual Property, the Law of Copyrights, Patents and Trademarks* (Thomson/West, St. Paul, Minn., 2003), p. 440; Govaere, Inge, *The Use and Abuse of Intellectual Property Rights in E.C. Law* (Sweet and Maxwell, London, 1996), pp. 14–15 and Lehmann, Michael, 'The Theory of Property Rights and the Protection of Intellectual and Industrial Property' (1985) 16 IIC 525, pp. 531–532.

and illustrates that IP law and antitrust law are closely intertwined in relation to tying. The analysis of the US patent misuse doctrine also illustrates that IP laws offer an interesting solution to the question of when to allow tying and when it should be unbundled.

In fact, tying is one of the most common allegations by infringers of patents and also the practice from which the patent misuse doctrine originates.[10] Other types of practice caught under the patent misuse doctrine include refusal to license, grant-back clauses,[11] price limitations or non-price resale restrictions, territorial limitations or field of use or customer limitations in licences,[12] discriminatory,[13] excessive and post-expiration royalties[14] and three more categories closely related to tying arrangements: tie-outs (or covenants not to deal in competing products or technologies),[15] package licensing,[16] and royalties based on total sale.[17]

2.   THE US PATENT MISUSE DOCTRINE

In the early 1900s, the US courts developed the doctrine of misuse. It originally only applied to patents, but was broadened in the early 1990s also to encompass copyright misuse.[18] The patent misuse doctrine is essentially a defence against claims of infringement of a patent; it is not a stand-alone claim. This means that the doctrine in principle only benefits infringers and is thus limited in applicability; however, there is no requirement upon these

---

[10]   Hovenkamp, Herbert, Janis, Mark D., and Lemley, Mark A., *IP and Antitrust, An Analysis of Antitrust Principles Applied to Intellectual Property Law* (Aspen Law & Business, New York, 2007), pp. 3–12 and Coch, Nicholas and Chen, Heidi, 'Specific Practices that have been Challenged as Misuse' chapter 2 in ABA Section of Antitrust Law, *Intellectual Property Misuse; Licensing and Litigation* (ABA, Chicago, 2000), p. 38.

[11]   *Transparent-Wrap Mach. Corp. v Stokes & Smith Co.*, 329 U.S. 637 (1947).

[12]   *General Talking Pictures v Western Electric*, 304 U.S. 175 (1938).

[13]   *USM Corp. v SPS Technologies*, 694 F.2d 505 (7th Cir. 1982).

[14]   *Brulotte v Thys*, 374 U.S. 29 (1964).

[15]   *In re Recombinant DNA Tech. Patent & Contract Litig.*, 850 F. Supp. 769 (S.D. Ind. 1994).

[16]   *Automatic Radio Mfg v Hazeltine Research*, 339 U.S. 827 (1950) and *Zenith Radio Corp. v Hazeltine Research*, 396 U.S. 100 (1969).

[17]   Coch and Chen, *supra* n. 10, pp. 37–69.

[18]   *Lasercomb America, Inc. v Reynolds*, 911 F.2d 970 (4th Cir. 1990) (http://digital-law-online.info/cases/15PQ2D1846.htm (26 July 2005)), which applies the misuse doctrine to copyright. There is, however, a significant divergence between the two doctrines, as the copyright misuse doctrine has early on been clear about its intended aim and demonstrated to be a broader defence than antitrust violations: *Lasercomb* at 978 and Hovenkamp, Janis and Lemley, *supra* n. 10, pp. 3–46.

infringers to prove harm caused by the misuse; hence, the misuse defence can be applied by *any* infringer who finds that his patentee has misused the patent.[19] The misuse must, however, fall within either or both of the two following categories:

(a)   Improper attempts to extend the scope of the patent (or other IP rights).
(b)   Violations of the US antitrust laws.[20]

Category (b) signifies that this doctrine is closely connected with antitrust law and to some extent provides the element binding antitrust and IP law together. In particular, conduct caught under the misuse doctrine would in many cases also have violated the US antitrust laws, but importantly comes into play as a defence under IP law and is therefore strongly grounded in IP law and its policies.[21] The link between the two legal systems occurred soon after the passing of the Sherman Act in 1890 when patent infringers attempted to rely on the then new antitrust law to defend their conduct.[22] However, the courts initially rejected all such attempts,[23] allowing only actions which went beyond the scope of the patent itself to be open to the antitrust law. [24]

In *Henry v Dick*,[25] a tying case prior to the development of the patent misuse doctrine, the Supreme Court demonstrated extreme munificence towards patentees by allowing the defendant to profit from the tying of an unpatented product with his patented invention.[26] As a direct result of the case the Clayton Act (1914)[27] was introduced by Congress and the Supreme Court moved to overrule the case in *Motion Picture Patents*,[28] thereby making illegal the tying of

---

[19]   Lemley, Mark, 'The Economic Irrationality of the Patent Misuse Doctrine' (1990) 78 *California Law Review* 1599, p. 1610.

[20]   Gordon, George and Hoerner J. Robert, 'Overview and Historical Development of the Misuse Doctrine', chapter I in ABA Section of Antitrust Law, Intellectual Property Misuse; Licensing and Litigation (2000), American Bar Association, US, p. 1.

[21]   Hovenkamp, Janis, and Lemley, *supra* n. 10, p. 3–2.

[22]   Gordon and Hoerner, *supra* n. 20, p. 3, see cases such as *Strait v National Harrow Co.*, 51 F. 819, 820 (C.C.N.D.N.Y. 1892); *Brown Saddle Co. v Troxel*, 98 F. 620 (C.C. N. D. Ohio 1899); and *United States Fire Escape Counterbalance Co., v Joseph Halsted Co.* 195 F. 295 (N.D. Ill. 1912).

[23]   *United States Fire Escape.*

[24]   Ibid., at 299.

[25]   *Henry v A. B. Dick Co.*, 224 U.S. 1 (1912).

[26]   Ibid., at 32.

[27]   Clayton Act 1914,15 U.S.C. §12–27.

[28]   *Motion Picture Patents* at 517.

unpatented products to extend the scope of the patent. *Motion Picture Patents* together with *Carbice* and *Leitch* laid the foundations of the patent misuse doctrine.[29] The three cases show that the real interest of the Supreme Court was *the extension of the patent in relation to unpatented staple products*,[30] whether this was done contractually or through a special business method.[31] It was secondary, which form the extension had taken place in.

In *Morton Salt*, the Supreme Court further expanded its concern to the effects the tie would have upon the market:

> Where the patent is used as a means of restraining competition with the patentee's sale of an unpatented product, the successful prosecution of an infringement suit even against one who is not a competitor in such sale is a powerful aid to the maintenance of the attempted monopoly of the unpatented article, and is thus a contributing factor in thwarting the public policy underlying the grant of the patent.[32]

Two later cases further extended the application of the patent misuse doctrine, *Mercoid* and *Mercoid v Honeywell*, before Congress stepped in, revised the Patent Act and introduced section 271(d) of the Patent Act (hereafter section 271(d)).[33] These cases had expanded the doctrine to non-staple components, proving one step too far for Congress,[34] whereas previous cases such as *Morton Salt* in principle carved out a safe haven for tying non-staple products with patented products.

Section 271(d) resulted in a significantly more feeble patent misuse doctrine, and its application was also drastically reduced after the patent reform,[35] although courts were not restricted in using developed case law in respect of staple goods,[36] neither did it hinder the Supreme Court in stopping

---

[29]   Hovenkamp, Janis and Lemley, *supra* n. 10, pp. 3–5 and Gordon, and Hoerner, *supra* n. 20, p. 11.

[30]   Staple and non-staple goods will be discussed further below. The distinction derives from the Patent Act itself, 35 U.S.C. § 271(c). A staple product is a ready, non-infringeable component, whereas a non-staple product is specially made to fit the patented product or process: Hovenkamp, Janis and Lemley, *supra* n. 10, pp. 3–17.

[31]   *Motion Picture Patents*, at 505, 508 and 518, *Carbice*, at 32–33, and *Leitch*, at 463.

[32]   *Morton Salt* at 493.

[33]   Patent Act 1952, 35 U.S.C. §271 (d)(1)–(3) and see Burchfiel, Kenneth J., 'Patent Misuse and Antitrust Reform: 'Blessed Be the Tie?' (1991) 4 *Harvard Journal of Law and Technology* 1, p. 20 and Hovenkamp, Janis and Lemley, *supra* n. 10, pp. 3–5 for comments on this.

[34]   *Dawson Chem. Co. v Rohm & Haas Co.*, 448 U.S. 176 (1980), at 201 and Burchfiel, *supra* n. 33, p. 20.

[35]   Hovenkamp, Janis and Lemley, *supra* n. 10, pp. 3–6.

[36]   Such as *Morton Salt* and *Motion Picture Patents*. In *Dawson* the Supreme Court acknowledged that *Mercoid* had been overruled by Congress' amendment, and

tying arrangements of IP-protected products with unprotected (non-staple) products; it merely did so under the antitrust rules instead.[37] In the Patent Misuse Reform Act of 1988 Congress made yet another statutory change, adding two more restrictions to section 271(d).[38] Section 271(d) states:

(d)   No patent owner otherwise entitled to relief for infringement or contributory infringement of a patent shall be denied relief or deemed guilty of misuse or illegal extension of the patent right by reason of his having done one or more of the following:

(1)   derived revenue from acts which if performed by another without his consent would constitute contributory infringement of the patent;

(2)   licensed or authorized another to perform acts which if performed without his consent would constitute contributory infringement of the patent;

(3)   sought to enforce his patent rights against infringement or contributory infringement;

(4)   refused to license or use any rights to the patent; or

(5)   conditioned the license of any rights to the patent or the sale of the patented product on the acquisition of a license to rights in another patent or purchase of a separate product, unless, in view of the circumstances, the patent owner has market power in the relevant market for the *patent or patented product* on which the license or sale is conditioned (emphasis added).

The changes are partly codifications of well-established principles derived from case law and partly controversial deviations from previous case law under the misuse doctrine.[39] Section 271(d)(5) in principle returned to the legal situation prior to *Motion Picture Patents*, and thus importantly allows patentees to tie unpatented staple products with their patented technology.[40] The important distinction developed in the misuse doctrine case law between un-patented staple and non-staple products and contributory infringement looks as if it has been forgotten by the Congress with the introduction of section 271(d)(5).[41] It therefore is at odds with previous Supreme Court case law, but less so with that of the lower courts such as the Court of Appeals in *Morton Salt*,[42]

---

thus the misuse doctrine was limited to tying of unpatented staple components: *Dawson Chem. Co. v Rohm & Haas Co.*, 448 U.S. 176 (1980), at 213–215 and Burchfiel, *supra* n. 33, p. 20.

[37]   *Fortner Enterprises, Inc. v United States Steel Corp. (Fortner I)*, 394 U.S. 495, 89 S.Ct. 1252, 22 L.Ed.2d 495 (1969) and *United States v Loew's Inc*, 371 U.S. 38, 83 S. Ct. 97, 9 L.Ed.2d 11 (1962) (albeit about copyright).

[38]   35 U.S.C. § 271(d)(4)–(5).

[39]   Burchfiel, *supra* n. 33, p. 6.

[40]   Ibid., p. 9.

[41]   Ibid., p. 10.

[42]   *G.S. Suppiger Co. v Morton Salt Co.*, 117 F.2d 968, 48 U.S.P.Q. 277 (7th Cir. 1941).

Mercoid,[43] and *Mercoid v Honeywell*.[44] In those rulings, the Court of Appeals had either applied antitrust law or looked upon the tie as a form of technological integration. That approach corresponded better with antitrust thinking and thus better with section 271(d)(5), which introduces an antitrust element, market power, to the doctrine. Finally, it could be argued that all Congress did was to create the missing link between antitrust law and the misuse doctrine, which the Supreme Court had acknowledged in several cases, but only applied comparatively.[45] Section 271(d)(5) acknowledges that there is a distinction between the patent and the products the patent protects, and that they are in principle in separate markets. The patent owner can, though, be found to have market power in either market for the exemption to kick in. This makes the exemption narrower than if it had market power only in the patented product market.

Burchfiel observes that section 271(d)(5) was not actually adopted due to concerns directly related to patent misuses or ties, but induced by the computer industry, which wanted statutory immunity to tie patented hardware with copyrighted software.[46] The amendment was later supported, not surprisingly, by a series of important Federal Circuit Court cases in the late 1990s in which it was made considerably harder to apply the misuse doctrine successfully.[47]

Peritz also endorses the amendment, although he warns:

[section 271(d)(5)] reflects good policy if the tie can be justified as an ex ante incentive to innovation which would not otherwise occur. In the absence of such justification, the doctrines in [section 271(d)] would reflect an ill-advised extension of property rights and restraint of competition that injure the public interest in promoting innovation.[48]

In other words, if section 271(d) is not interpreted strictly, the approach to extension of patents may revert to the days of *Henry v Dick*, the kind of case the patent misuse doctrine was introduced to catch.

---

43   *Mercoid* at 664–665.
44   *Mercoid v Honeywell*, at 683.
45   *Motion Picture Patents*, at 517, *Carbice* at 32–34, and *Morton Salt* at 490.
46   Ironically enough no equivalent amendment to the laws on copyright was made by Congress: Burchfiel, supra n. 33, pp. 21–22.
47   *In re ISO Antitrust Litig.*, 203 F.3d 1322 (Fed. Cir. 2000); *Virginia Panel Corp. v MAC Panel Corp.*, 133 F.3d 860 (Fed. Cir. 1997); *Braun Med. Inc. v Abbott Labs*, 124 F.3d 1419 (Fed. Cir. 1997).
48   Peritz, R. J. R., 'Competition Policy and Its Implications for Intellectual Property Rights in the United States', chapter 3 in Anderman, Steve (ed), *The Interface between Intellectual Property Rights and Competition Policy* (Cambridge University Press, Cambridge, 2007), p. 142.

Section 271(d)(5) illustrates three important policies of Congress. First, in relation to the balance between the protection of private property/private innovation and placing novel information in the public domain, Congress favours the former. Second, Congress does not believe that IP automatically confers market power, at least not significant market power, otherwise section 271(d)(5) would make no sense.[49] Third, from a tying perspective, this offers a glimmer of hope – recognition by Congress that tying should not always be seen as illegal *per se*. In fact, section 271(d)(5) makes tying *legal per se*, and only when the particular conditions are present will the tying violate the patent laws.

The Patent Act amendments leave a question mark over the purpose of the misuse doctrine and its continuation, as well as its application.[50] The doctrine's downfall seems to be the many cases which could equally well have been dealt with under antitrust law. Interestingly, the patent misuse doctrine cases have all acknowledged the importance of antitrust law, but avoided its direct application. In *Mercoid*, the Supreme Court attempted to create a dividing line between the two, and it further dwelt upon this in *Mercoid v Honeywell*, remarking that attempts to bring unpatented products within the ambit of the exclusive right a patent grants is appraised not by IP law, but antitrust law. [51]

In the eyes of many people, the patent misuse doctrine has become superfluous, while such people forget that the doctrine in principle embraces more wrongs than just anti-competitive behaviour.[52] Moreover, the equity origin often seems to be overlooked when the application of the misuse doctrine is debated, and in particular its connection to antitrust law.[53] It is, however, necessary to keep in mind that the patent misuse doctrine is not only about protecting competitors, but also about ensuring that the true purpose and high standard of the patent system are upheld. Therefore, other actions by patent

---

[49]   Confirmed by the Supreme Court in *Illinois Tool Work Inc. v Independent Ink Inc.*, 547 U.S. 1 (2006), at 12; compare with Burchfiel writing prior to *Illinois Tool*: Burchfiel, *supra* n. 33, p. 23 and Coch and Chen, *supra* n. 10, n. 12.

[50]   See comments by Lemley (1990), *supra* n. 19 and Burchfiel, *supra* n. 33.

[51]   *Mercoid v Honeywell* at 684.

[52]   *C.R Bard, Inc. v M3 Systems*, 157 F.3d 1340 (Fed. Cir. 1998) and *USM Corp. v SPS Technology, Inc.*, 694 F.2d 505 (7th Cir. 1982); see also Burchfiel, *supra* n 33; Hovenkamp, Janis and Lemley, *supra* n. 10; Lemley (1990), *supra* n. 19; and Merges, Robert, 'Reflections on Current Legislation Affecting Patent Misuse' (1988) 70 *J. Pat. & Trademark Off. Soc'y* 793.

[53]   *Morton Salt*, at 493; see also Gordon, and Hoerner, *supra* n. 20, p. 14; *United States Gypsum Co. v National Gypsum Co.*, 352 U.S. 457 (1957), at 465 where the Court states: 'The [patent misuse doctrine] is an extension of the equitable doctrine of "unclean hands" to the patent field'.

owners, which do not have an anti-competitive effect but which nevertheless seek to extend a patent beyond its rightful boundaries, will also be caught under the doctrine. An attempt to monopolise unpatented products is subordinate to this.

As this chapter is not intended to be a discussion of the existence of the misuse doctrine, we will leave this issue aside, but merely note that the amendments have significantly limited the misuse doctrine, and thus it is less of a powerful defence for patent infringements.[54] That said, it is still available as a defence for patent infringement when the conditions in section 271(d) are met. For tying, that means when the patentee has extended his patent to an unpatented (staple or non-staple) product by tying it to the patented product or process. Moreover, the patentee must hold market power either in the market for the patent or the tying (patented) product market and the patent itself cannot be applied to presume market power.[55] Therefore, section 271(d), and in particular section 271(d)(5), is a positive development for tying arrangements as it recognises that in certain circumstances tying is not anti-competitive and can be beneficial and therefore permissible, even if it means that the exclusive right of the patent has been extended to encompass an unpatented component, as will be discussed below.

3.   TYING ARRANGEMENTS CAUGHT UNDER THE PATENT MISUSE DOCTRINE

The limiting of patent misuse to situations where the patentee holds market power by the introduction of section 271(d)(5) has caused the rule to become an intermediate standard between the previous *per se* rule and current antitrust law – in fact, it works as a form of bright line rule, creating a safe haven for certain conduct when there is no market power and more importantly sees tying as *legal per se*.[56] Section 271(d)(5) requires merely an assessment of market power. One cannot rely on the assumption that the patent itself gives

---

54   For discussions of the misuse doctrine's future see Judge Posner in *USM Corp v SPS Technology*; Burchfiel, *supra* n. 33; Gordon and Hoerner, *supra* n. 20; Hovenkamp, Janis and Lemley, *supra* n. 10, chapter 3; Lemley (1990), *supra* n. 19; Merges, *supra* n. 52; Calkins, 'Patent Law: The Impact of the 1988 Patent Misuse Reform Act and Noerr-Pennington Doctrine on Misuse Defenses and antitrust Counterclaims' (1989) 38 *Drake Law Review* 175, and Webb, Jere M. and Locke, Lawrence A., 'Intellectual Property Misuse: Developments in the Misuse Doctrine', (1991) 4 *Harvard Journal of Law and Technology* 257.

55   *Illinois Tool* at 12 and 16.

56   *Virginia Panel* at 869.

the patentee market power, although scholars have expressed such concerns and this is indeed plausible if section 271(d)(5) is read literally.[57] Although such an assumption is old-fashioned the wording of section 271(d)(5) itself does little to pacify such fears.[58]

That said, the Supreme Court in *Illinois Tool* made clear that it understood section 271(d)(5) as meaning that the mere existence of a patent did not constitute the requisite market power.[59] This ruling establishes that the presumption of market power due to a patent is no longer sufficient. The plaintiff must give evidence which goes beyond the mere ownership of the patent as proof of market power. However, it does not mean that a patentee cannot even be seen to hold market power due to its patent. It may still be the case – evidence for it must be provided by the plaintiff. Hovenkamp *et al.* believe that the market power requirement will additionally necessitate a fully-fledged economic inquiry of the market and barriers to entry.[60] Section 271(d)(5) appears silent on the matter, but the wording 'in view of the circumstance' could be interpreted as endorsing such assessment. In *Virginia Panel* the Court indicated that a court should look to the effect of the market power and the expansion of the exclusive right by tying; if this unduly harms competition and distorts the tied product market, it is caught by the section.[61]

Crucially for the patent misuse doctrine, *Virginia Panel* clarified that in the absent of market power, tying is no longer illegal *per se*, because of section 271(d)(5). The Court noted that where it is unclear whether the conduct falls under *per se* illegality or is specially exempted under section 271(d), the court must assess under a *rule of reason* approach whether the conduct falls reasonably within the scope of the patent (that is 'it relates to subject matter within the scope of the patent claims'[62]) or has the effect of extending the scope of the patent and 'does so with an anti-competitive effect'.[63] This last point implies that even if the conduct goes beyond the scope of the patent grant, it

---

57   Coch and Chen, *supra* n. 10, pp. 39 and 42, referring to early cases of the patent misuse doctrine: 'necessary [market] power was assumed from the presence of the patent on the tying product'. See also Antitrust Guidelines for the Licensing of Intellectual Property, issued by the US Department of Justice and the Federal Trade Commission , 5 April 1995 (www.usdoj.gov/atr/public/guidelines/0558.htm), Section 2.2; *Jefferson Parish Hosp. Dist. v Hyde*, 466 U.S. 2 (1984), at 16 and *Illinois Tool*, at 12 and 16.
58   See Burchfiel, supra n. 33, p. 23, but note *Illinois Tool*, at 12 and 16.
59   Ibid., at 12.
60   Hovenkamp, Janis and Lemley, *supra* n. 10, p. 3–11.
61   *Virginia Panel* at 869 and Hovenkamp, Janis and Lemley, *supra* n. 10, p. 3–8.
62   *Virginia Panel*, at 869 and see also *Mallinckrodt, Inc. v Mediapart, Inc.*, 976 F. 2d 700 (Fed. Cir. 1992) at 708.
63   *Virginia Panel*, at 869.

may not have a sufficient anti-competitive effect and thus will not necessarily be caught by the doctrine.[64]

This means that the *rule of reason* assessment requires an analysis of the pro-competitive as well as the anti-competitive effects triggered by the practice undertaken by the patentee. Under this assessment, the patentee will have the opportunity to justify its conduct. Such justification must have a base in sound economics or be market-related.[65] *Virginia Panel* therefore aligns case law with the antitrust rules on tying, in comparison to the statute, which is more or less silent on the matter, although importantly the antitrust rule on tying sees tying as illegal *per* se. It also means that *Virginia Panel* can be interpreted as having limited the originally broad scope of the patent misuse doctrine to mere antitrust violations[66] by introducing an assessment of anti-competitive effects.[67]

On the other hand, it is questionable whether the legality of the tying arrangement can be adequately assessed without some investigation into the effects upon the tied product market. Hovenkamp *et al.* are certainly of this opinion, although their conclusion is based upon antitrust case law.[68] The introduction of an anti-competitive effects assessment under patent misuse would, as noted above, align the approach with that of antitrust and thereby avoid rule shopping. The courts' eagerness to do so indicates that in their opinion tying arrangements are perhaps better dealt with under antitrust rules.[69]

In *Illinois Tool*, as noted in Chapter 4, the Supreme Court held that the tying of patented products should be assessed by the standards adopted in antitrust cases *Fortner II*[70] and *Jefferson Parish*, and not the *per se* rules applied in *Morton Salt* and *Loew's*.[71] It continued by acknowledging that some tying arrangement will remain unlawful as a result of a true monopoly or a marketwide conspiracy,[72] but such conclusion must be reached on the basis of a market power assessment rather than a mere assumption.[73] This statement could mean the true end of the misuse doctrine applied to tying arrangements;

---

64   Gordon, and Hoerner, *supra* n. 20, p. 27.
65   Hovemkamp, Janis and Lemley, *supra* n. 10, pp. 3–11.
66   *C.R. Bard v M3* at 1372.
67   *Virginia Panel* at 868.
68   Hovenkamp, Janis and Lemley, supra n. 10, pp. 3–15–16 and see *Northern Pacific Railway Co et al. v United States*, 356 U.S. 1,2 (1958).
69   *Virginia Panel* at 869 and *Illinios Tool*, at 12.
70   *United States Steel Corp. v. Fortner Enterprises, Inc.*, 429 U.S. 610, 622 (*Fortner II*).
71   *Illinois Tool* at 13.
72   As was the case in *United States v. Paramount Pictures, Inc.*, 334 U.S. 131, 145–146 (1948).
73   *Illinois Tool* at 13.

but importantly, it also paves the way for a more flexible approach to tying arrangements in general.

Currently the following conditions must be established for the tying arrangement to be illegal under the patent misuse doctrine:

(1) two separate products,
(2) the licensing or sale of the patented product is conditional upon the purchase or licensing of another (patent or) product, and
(3) the licensor or seller has market power.[74]

As regards whether there are two separate products, the test is based on the physical appearance of the products rather than the consumer demand test found under antitrust law.[75] Prior to section 271(d), a distinction was made between staple and non-staple products.[76] It is unclear whether section 271(d)(5) was intended also to exempt the tying of unpatented staple products; it appears in fact to be all-inclusive.[77] *Virginia Panel* is rather vague on the matter, and merely notes that case law held the tying of staple products to be illegal *per se*, whereas Congress carved out an exemption for all tying arrangements.[78] Hovenkamp *et al.* have also not addressed this issue in their otherwise lengthy and detailed analysis of the misuse doctrine.[79] A strict reading of section 271(d)(5) indicates that the exemption indeed covers *all* tying arrangements, and therefore the previous staple/non-staple distinction has become redundant. That said, there may be certain circumstances in which the distinction can come into play, in particular, when the *rule of reason* approach is applied, when it is unclear whether the patented product confers market power. In those situations the Court in *Virginia Panel* held that it was necessary to look to whether the conduct felt reasonably within the scope of the patent, and especially whether it related to the subject matter of the patent.[80] The language of previous case law implies that this refers to the staple/non-staple distinction. The tied product is staple if it is within a competitive market and not closely related or specially adapted to the invention,[81] although it is uncertain whether these are cumulative requirements or requirements which can be applied individually. On the other hand, if the product is non-staple, that is the product has

---

74   Coch and Chen, *supra* n. 7, p. 39.
75   See *Jefferson Parish*, at 21–25 and Coch and Chen, *supra* n. 7, pp. 39–40.
76   *Senza-Gel Corp. v Seiffhart*, 803 F.2d 661 at 663 (Fed. Cir. 1986).
77   Burchfiel, *supra* n. 33, p. 10.
78   *Virginia Panel* at 869.
79   See Hovenkamp, Janis and Lemley, *supra* n. 10, chapter 3.
80   *Virginia Panel* at 869.
81   Coch and Chen, *supra* n. 10, p. 39.

no other use than as part of the patented device or combination, the patentee may tie it together with the patented part of the product. [82]

There is no similar distinction between staple and non-staple products under antitrust law as such; however, a broad reading of non-staple products can be compared to technological integration. If such a reading is accepted, then the exemption of non-staple product makes sense, as technological integration has received more lenient treatment under antitrust law.[83] From an economic perspective it is worth recalling that economic theories suggest that it is not possible to extract two monopoly profits from complementary products applied in fixed proportions,[84] but an additional caveat is added by requiring that the market for the tied product to be competitive.[85] In other words, if there is already a competitive market for a commodity there should be limited risk of harm to competition. This perhaps explains the breadth of section 271(d)(5) to allow for the tying of staple products; but note there is no limitation upon the exemption if the tied product market is not competitive. Moreover, if the tied product is non-staple, it is unlikely to be a competitive market. Therefore, the permission to tie non-staple unpatented products with the patented product makes sense only if there is no demand for the former product without the patented component: in other words if the products are technologically integrated.

From an aftermarket perspective the distinction between non-staple and staple products offers a more flexible outlook upon the products, because credit is awarded to the relationship between the combined products. As was demonstrated in Chapter 3, unlike the US patent misuse doctrine, EC competition law does not allow for a distinction between staple and non-staple goods when there are IP rights involved. This means that the EC competition law approach appears inflexible and strict, and consequently can have a damaging effect on innovation because the relationship between the combined products will not be given sufficient credit. It is therefore worth considering the intro-

---

[82]   Burchfiel, *supra* n. 33, p. 9.

[83]   See *US v Microsoft Corp.*, 147 F.3d 935 (D.C. Cir 1998) (*US Microsoft II*) and *US v Microsoft Corp.* 253 F.3d 34 (D.C. Cir. 2001) (*US Microsoft III*).

[84]   Whinston, Michael D., 'Tying, Foreclosure, and Exclusion' (1990) 80 *The American Economic Review*, 837, Whinston, Michael D., 'Exclusivity and Tying in *U.S. v. Microsoft:* What We Know, and Don't Know' (2001) 15(2) *Journal of Economic Perspectives*, 63, p. 70'; Bishop, Simon and Walker, Mike, *The Economics of EC Competition Law, Concepts, Application and Measurements* (Sweet and Maxwell, London, 2002), paras. 6.63; and Jacobs, Michael, 'Third Line Forcing and Tying Arrangements – Some Comments on the United States' Position', article published on Blake Dawson Waldron web site: Competition and Consumer Protection (www.bdw.com.au/areas/tradespractices/tpa-new-4.htm).

[85]   Bishop and Walker, *supra* n. 84, paras. 6.63 and Jacobs, *supra* n. 84.

duction of a similar approach to products under EC competition law in relation to tying in aftermarket products.

In comparison with US antitrust, the elements that are required here for the establishment of an illegal tying arrangement are very similar to the ones found under those rules. It is important to note the absence of serious differences between the two legal doctrines in respect of tying. Both require the establishment of two separate products, albeit that the actual analysis of this may differ and both require evidence of coercion and market power. Case law dictates that both approaches should include an anti-competitive effects analysis, which will include an assessment of any objective justifications for the tying arrangement. [86] However, the misuse doctrine's starting point is from a *legal per se* perspective compared to the illegal *per se* antitrust approach. This has an important impact upon attitudes in tying litigation. The antitrust rules require some evidence of harm and anti-competitive effects which is not required under section 271(d)(5), although *Virginia Panel* indicates that such assessment may be necessary if the tie is not found to be legal.

Finally, from an economic perspective section 271(d)(5) is interesting as it coincides with some of the recommendations made in the economics chapter. First, it allows tying without market power to be *legal per se*, thereby creating a safe haven. Second, it calls for an assessment of market power, which was also one of the important points stressed in the economics chapter – tying is unlikely to cause harm unless the company in question holds significant market power in the tying product market, and *Illinios Tool* made it clear that the mere presence of a patent does not fulfil this condition. Third, *Virginia Panel* allows for a pro- and anti-competitive effects analysis, in line with economic findings when the conduct falls under the *rule of reason*; in other words, when the tying falls outside the safe haven. Although section 271(d)(5) is rather concise and therefore does not fully correspond to the recommendations identified in the economics chapter, there are many similarities, and perhaps its approach offers a realistic starting point for a more modern test of tying under antitrust law.

4. THE US MISUSE DOCTRINE AND ITS EFFECT ON TYING – LESSONS TO BE LEARNED

The Supreme Court has developed an aggressive approach to tying arrangements under both antitrust law, where a tying arrangement is regarded as more often than not illegal *per se* with a presumption of market power, and the

---

[86] See *Yentsch v Texaco, Inc.*, 630 F. 2d 46 (2nd Cir. 1980).

patent misuse doctrine, which condemned some ties of patented products which could not even have fulfilled the low antitrust standards.[87] Both approaches disregard the numerous benefits tying arrangements can bring.

Although applying a structurally similar test, the current rules under US patent law have a more positive look upon tying arrangements involving patented goods accepting them as *legal per se and places the onus upon the infringer to establish market power.*

This chapter teaches us two essential points about the approach to tying. Firstly, in the US after a significant lapse of time it has been left almost entirely to competition law to deal with the conduct of tying. Secondly, the US Patent legislation has recognised that tying becomes an issue only once the patent owner has market power – this approach corresponds well to the antitrust approach as well as economic theory. Moreover, involving market power as an assessment tool indirectly throws the ball back into the antitrust law corner, and this has further been emphasised by case law calling for an anti-competitive effects assessment of the tying arrangement.

Although there has been a battle between IP laws and antitrust laws regarding which is best suited to dealing with tying arrangements involving IP rights, the approaches to tying within each field have been remarkably similar. That said, section 271(d)(5) has carved out a very important exemption for tying arrangements which, although appearing similar in construction to antitrust rules, marks an important step towards a more flexible approach to tying, in particular the fact that tying is *legal per se*, until market power has been proven. Section 271(d)(5) therefore offers a good initial alternative to the current antitrust approach, although it is rather vague in respect of the so-called in-between cases. Case law such as *Virgina Panel* has been quick to introduce a *rule of reason* approach in these situations. Again, it must be said to be a more favourable attitude to tying than the stricter *per se* illegality under the antitrust rules. An alignment of the antitrust rules and IP laws with tying is naturally welcomed to ensure legal certainty, but the approach under IP law should prevail as this in fact offers a much more modern outlook upon tying – one which is in line with economic thinking as well.

---

[87]   Hovenkamp, Janis and Lemley, *supra* n. 10, pp. 21–4.

6. Alternative Solutions, the Development of a New Regulatory Model

1. INTRODUCTION

In this chapter, we will re-assess the five steps approach applied to tying in an attempt to identify and develop a new and better method to deal with tying. Such a new approach must take into account the need for greater flexibility to ensure that all anti-competitive tying arrangements are caught, but the pro-competitive ties escape. Therefore the approach must be more flexible towards the company, the product(s) and the market conditions in which the company operates. The approach sought needs to be in line with the current trend in economic thinking, but must also consider the review of Article 82 and Section 2, and the calls made there for a more economic approach. There are therefore several points which need to be considered.

First, although an obvious question with an obvious answer: should the current test stay? Does it sufficiently meet the competition requirements for modern day industries? It goes without saying that the answer to that question in the eyes of the author is a clear-cut no. Chapters 2, 3 and 4 have clearly demonstrated that the current test is flawed and ill-equipped to deal with high-technology industries, rapid innovation and product development. We must therefore re-think and re-evaluate the test to identify another more suitable solution.

This leads to the second point: is the approach to tying arrangements the correct one? Should we maintain an illegal *per se* approach to tying or should we move towards a *legal per se* approach instead or somewhere in between? Again the previous chapters have established that the illegal *per se* approach is too strict and does not adequately give weight to the benefits tying can generate. However, the previous chapters have not clearly answered where we should move. A fuller discussion of this question will therefore be dealt with in section 2 below, and a recommendation will be made.

The third question arising is should the format of the test remain the same? Section 3 will discuss this question in greater detail. However, one of the points which has been raised both in the Article 82 review debate and by the

DoJ in its *Section 2 Report*,[1] is whether this is a quest for an one-rule-fits-all which captures and treats all anti-competitive conduct the same, or whether a separate, yet more economic rule for different types of anti-competitive conduct should be sought. Whichever approach is taken, and whichever rule or rules are adopted, they must be broad enough to capture all anti-competitive conduct, yet sophisticated enough to leave out pro-competitive behaviour. Regardless of the choice of rule, the rule adopted must be sufficiently broad to capture all exclusionary conduct, but also sufficiently fine-tuned to reveal benefits and efficiencies, and identify when the behaviour is pro-competitive. In the opinion of the author, it is possible to apply an all-encompassing rule for the various different types of conduct, but some screening must take place beforehand to ensure that the pro-competitive effects of a particular behaviour are given sufficient credit. The first screen is naturally the question of dominance, which is discussed in section 5, but for tying there is an option for another screen, that of the need for two separate products. Section 4 of this chapter will discuss this latter screen and question the need for it. A tying arrangement is only harmful if it causes (current or future) anti-competitive effects; section 6 will therefore assess different options for a new anti-competitive effects test. The anti-competitive effects discussion, however, is not complete without an assessment of the objective justifications, the assessment which in principle brings to light the pro-competitive effects of the behaviour and is therefore crucial to any assessment of exclusionary behaviour. The assessment of objective justifications will be discussed in section 7.

Each section contains several options or alternative solutions for a new approach and will conclude with a recommendation of the best test applicable to each step in the approach to tying, taking into consideration the above-mentioned points and requirements a change will need to fulfil. These recommendations will be combined at the end of the chapter to create the new suggested approach to tying.

2. ILLEGAL *PER SE* – *RULE OF REASON* – WHERE ARE WE NOW? WHERE SHOULD WE BE?

The above chapters highlighted several interesting points about the similarities

---

[1]   US Department of Justice, 'Competition and Monopoly: Single-Firm Conduct Under Section 2 of the Sherman Act' (2008) (www.usdoj.gov/atr/public/reports/236681.htm) (*Section 2 Report*). Note the report was controversially withdrawn by the DoJ in May 2009, see press release 04-459 'Justice Department Withdraws Report on Antitrust Monopoly Law' 11 May 2009 (www.usdoj.gov/opa/pr/2009/May/09-at459.html).

and differences between the US and EC approaches to tying arrangements. The assessment of tying arrangements under IP law identified a very positive drive towards the legality *per se* of tying arrangements under US law. This push has occurred alongside antitrust law and, in fact, the two approaches – that of US IP law and that of US antitrust law – supplement and support each other. *Illinois Tool* is a testament to that effect.[2] There is, however, one significant difference, and that is that section 271(d)(5) of the US Patent Act clearly sees tying arrangements involving patents as legal *per se* and assesses tying arrangements from this perspective, whereas the antitrust approach has traditionally assessed tying arrangements as illegal *per se*. The recent cases such as *Jefferson Parish*[3] and *US Microsoft II*[4] and *III*[5] have indicated a change in that attitude, requesting Microsoft's product integration to be assessed under a *rule of reason* approach.[6] If this line is upheld, it may be the first step towards a legal *per se* approach.[7] Equally, the *Section 2 Report* recommends that tying should not be judged under the *per se* illegality rule.[8]

In comparison, the EC has maintained a stable attitude towards tying arrangements – that of (almost) illegality *per se*; almost, because of the requirement of market power– as is confirmed by the recent *EC Microsoft* case. The CFI took a very legalistic approach to tying arrangements, and in particular to the analysis of the potential anti-competitive effects the tie could have. It thereby foreclosed the opportunity for moving the Article 82 analysis towards a more economic approach, which the Commission had initiated with its anti-competitive effects assessment in *EC Microsoft*.

The EC IP law approach to tying arrangements has been dominated by competition law to such an extent that in some countries IP law itself does not assess tying arrangements involving IP rights.[9] These are now only assessed through competition law.

These differences in approach across the Atlantic have occurred despite the fact that both jurisdictions have very similar sets of conditions which must be present for a finding that the tying arrangement is illegal. In particular certain differences seem to have made a big impact, such as the understanding of consumer demand and product innovation.

---

[2]   *Illinois Tool Work Inc. v Independent Ink Inc.*, 547 U.S. 1 (2006).
[3]   *Jefferson Parish Hosp. Dist. v Hyde*, 466 U.S. 2 (1984).
[4]   *US v Microsoft Corp.*, 147 F.3d 935 (D.C. Cir 1998) (*US Microsoft II*).
[5]   *US v Microsoft Corp.*, 253 f.3d 34 (D.C. Cir. 2001)(*US Microsoft III*).
[6]   Ibid., at 84.
[7]   Bellis, Jean-François, 'The Microsoft Judgment' Conference, 25 September 2007, BIICL, London.
[8]   *Section 2 Report*, *supra* n. 1, pp. 89–90.
[9]   UK Patent Act 1977, section 44, repealed in March 2000 by section 70 of the UK Competition Act 1998.

The economic findings in Chapter 2 and the analysis of case law illustrated that many improvements could be made to the set of conditions applied to tying arrangements to ensure that harmless tying arrangements are not caught under the strict rules currently applicable.

The first question to answer, though, is how tying should be perceived overall. The findings above indicate that there should be a move away from the current illegal *per se* approach. Below some alternative solutions are assessed in this respect.

2.1. Good or Bad Conduct

Eilmansberger designs a theory on how to identify good competition from bad, and in order to do so, he categorises abusive behaviour into two groups: one form of behaviour is behaviour that he identifies as abusive in itself and the other form is that which is abusive due to its effect.[10] The consequence of this distinction is that the first group consists of behaviour which, according to Eilmansberger, should be seen as illegal *per se*; the other group is less dangerous and can therefore be assessed under a *rule of reason* approach.[11]

The reason for highlighting Eilmansberger's theory is his opinion about tying arrangements, as he has categorised these under the first group and therefore argues that tying should be illegal *per se* (when the company is dominant).[12] This is because the market power abuses such as refusal to supply or tying, where the leveraging of market power is used to gain access to and success in a downstream market instead of superior business performance, are clearly dominant specific distortions of competition.[13] Therefore, although an appreciability threshold can be applied in theory, in practice Eilmansberger argues that this makes little sense, especially considering that the conduct is so obviously abusive.[14]

Eilmansberger's view can in principle be accepted because leveraging of market power clearly is something achievable only with sufficient dominance, if not a near monopoly, in a primary market. It does not follow, however, that all forms of tying conducted by dominant firms are of an abusive nature. If the tying commenced long before the company gained a dominant position and

---

[10]   Eilmansberger, Thomas, 'How to Distinguish Good from Bad Competition Under Article 82EC: In Search of Clearer and more Coherent Standards for Anti-competitive Abuses' (2005) 42 *CMLRev* 129, p. 143.

[11]   Ibid.

[12]   Ibid.

[13]   Ibid.

[14]   Ibid., p. 152.

the tied products are now seen as one or integrated,[15] or if it was part of commercial usage, tying should not be deemed illegal, at least without significant proof of the secondary market being adversely affected. Eilmansberger's theory will also have severe consequences for IP rights owners, who have often been seen as holding a *de facto* monopoly, without a change in how the relevant market is defined. Moreover, his theory does not take into consideration that there can in fact be pro-competitive effects of tying even if conducted by a dominant company, and thus even in a case where there are no likely or foreseeable foreclosure effects, there could still be a finding of abuse.

Current tying case law has shown that abuse or coercion cannot be established before anti-competitive effects have been identified. If it is, then the Eilmansberger theory above will apply and tying becomes illegal *per se*. There is nothing in the wording of Article 82 to suggest such an interpretation.[16] The current model applied to tying requires proof of coercion and foreclosure effects, but allows for objective justifications. Although more or less disregarded by the CFI, the Commission embraced a standard of proof of effects in *EC Microsoft*.[17]

Eilmansberger's theory of labelling tying as bad without considering its effects is therefore an old-fashioned form-based approach and inconsistent with the current trend towards a more effect-based approach. It is also inconsistent with the development under US case law, where courts, despite applying a *per se* illegality rule to tying, introduced several conditions including an assessment of the effects upon the tied product market before finding a tying arrangement illegal. It is therefore clear from the evidence gathered so far that illegal tying should not be concluded without assessing the effects of the tie upon the markets, a strict *per se* illegality rule is therefore not appropriate and it must be dismissed.

2.2. *Rule of Reason*

The above conclusion leads us to another form of test at the other end of the scale to that of *per se* illegality. O'Donoghue and Padilla correctly note that for tying or other forms of leveraging abuse there must be evidence of abusive behaviour in the form of coercion, that is making the purchase of one product conditional upon the purchase of another or refusing benefits such as reductions

---

15 See argument by Microsoft, in Case COMP/ C-3/37.792 *Microsoft*, [2005] 4 CMLR 965, paras. 814–20.
16 The economic analysis identified that the illegal *per se* approach does not reflect the economic dynamics surrounding tying: Chapter 2, Section 7.
17 *EC Microsoft*, Commission Decision, Section 5.3.2.1.4, and Case T–201/04 *Microsoft Corp. v Commission* [2007] ECR II-3601, para. 1090.

in rebates when the products are purchased separately.[18] The mere use of dominance in one market to gain an advantage in another market is not sufficient to establish tying abuse.[19] In most cases O'Donoghue and Padilla's coercion condition has been fulfilled.[20] However, in *EC Microsoft*, although there was some coercion of the OEMs having to ship computers containing Windows with Windows Media Player (WMP) in the form of licensing agreements requiring this inclusion in the shipment, there were still alternative means of obtaining the WMP, the tied product and other media players. Importantly, the pre-installation of WMP did not force users of Windows to make use of WMP or hinder them in purchasing or downloading for free other media players onto Windows.[21] O'Donoghue and Padilla's test is thus not fulfilled in *EC Microsoft*.

It is suggested that coercion or tying should not be concluded separately but in conjunction with the assessment of anti-competitive effects. This is reinforced by the need for a causal link between the dominance and the abusive behaviour. In particular, for tying where two markets are involved it is important that the abusive behaviour in the tied product market is related to the dominance in the tying product market. However, this cannot be established until it is proven that there are indeed anti-competitive effects in the tied product market. The question left open is the level of standard of proof. Should the anti-competitive effects already be present in the tied product market? Or should there be a foreseeable risk that anti-competitive effects will occur? O'Donoghue and Padilla argue that if there are no or limited anti-competitive effects found in the tied product market, the tying is merely use of dominance in the tying product market to gain an advantage in that market, and that should be acceptable.[22]

Anti-competitive effects and standard of proof will be discussed below, for now it suffices to say that a balancing test of the pro- and anti-competitive effects would indeed be a step in the right direction towards the effect-based approach. The difficulty with such an approach is the need to assess each case on its own facts, and this risks causing mayhem to legal certainty as well as being very costly. A full blown *rule of reason* approach to tying, is therefore

---

18   O'Donoghue, Robert and Padilla, A. Jorge, *The Law and Economics of Article 82 EC* (Hart Publishing, Oxford and Portland, Oregon, 2006), p. 210.

19   Ibid.

20   *International Salt Co. v United States*, 332 U.S. 392, 68 S.Ct. 12, 92 L. Ed. 20 (1947); *Northern Pacific Railway Co. et al. v United States*, 356 U.S. 1 (1958); *Hilti* Commission Decision; Case IV/30.787 *Eurofix-Bauco/ Hilti* [1988] OJ L65/19, para 75 and Case T–83/91 *Tetra Pak Rausing SA v Commission (Tetra Pak II)* [1990] ECR II–309, [1991] 4 CMLR 334, para. 137.

21   Case T–201/04 *EC Microsoft*, paras. 968–970.

22   O'Donoghue and Padilla, *supra* n. 18, p. 210.

not an advisable option. This leads to the conclusion that a middle-way solution should be sought, one which moves away from the rigid form-based approach and towards a more flexible and economic approach, albeit allowing for the screening of the conduct and company for specific features, which could quickly rule out severe anti-competitive concerns. In other words, a *rule of reason* approach must also be rejected.

2.3. *Per se* Legality

One such middle-way approach is, surprisingly, enough to view tying from a *legal per se* perspective. This would mean that tying would be accepted as a legitimate business strategy also when undertaken by a dominant company. This idea is not completely alien – in fact this is the approach applied under the US patent misuse doctrine and an approach advocated by some economists and lawyers.[23] This approach would give sufficient weight to the efficiencies and benefits that a tying arrangement generates as recognised by economists. However, it should be recalled that tying arrangements can at times be harmful and foreclose competition from the market, and therefore some restriction is essential to ensure that the harmful types of tying arrangements are caught. The solution applied under the US patent misuse doctrine is to introduce a screen. It essentially permits tying arrangements unless the company in question has significant market power in the tying product market. Under the current approach to tying on both sides of the Atlantic there is already a requirement for market power and, consequently, it is thus easy to introduce *per se* legality. This will be discussed below in section 5. However, the whole assessment of a tying arrangement depends in effect upon whether there are in fact two products. As a result, a screen should be included assessing this fact – if there are not two products there is no need to take the assessment any further. The inclusion of these two screens or safe havens is in line with the recommendations made by economists as identified in Chapter 2. It is also in

---

23   Alhborn, Christian, Evans, David S., and Padilla, A. Jorge, 'The Antitrust Economics of Tying: A Farewell to *Per Se* Illegality' [2004] *Antitrust Bulletin* 287, Evans, David S, and Salinger, Michael, 'Why Do Firms Bundle and Tie? Evidence from Competitive Markets and Implications for Tying Law' (2005) 22 *Yale Journal on Regulation* 37; Hylton, Keith and Salinger, Michael, 'Tying Law and Policy: A Decision-theoretic Approach' (2001) 69 *Antitrust Law Journal* 469; Lind, Robert and Muysert, Paul, 'Innovation and Competition Policy, Challenges for the New Millennium' (2003) 24 *European Competition Law Review* 87; O'Donoghue, Robert and Padilla, A. Jorge, *supra* n. 18; and Ridyard, Derek, 'Tying and Bundling – Cause for Complaint?' (2005) 26 *European Competition Law Review* 316; Hovenkamp, H., 'IP ties and Microsoft's rule of reason' [2002] *The Antitrust Bulletin* 369 and Bellis, *supra* n. 7).

line with previous case law, albeit that the approach to these screens comes from a different angle – that of legality – and therefore offers a much more positive outlook upon tying and creates greater legal certainty for companies. The *per se* legality approach permits the pro-competitive tying arrangements to flourish, whereas the proposed screens will allow the competition authorities to curb and control harmful ties. The *per se* legality approach is therefore recommended.

2.4.   Recommending *per se* Legality

The three assessments above discussed the approach to tying. The first approach, that of *per se* illegality, was dismissed for being old-fashioned and out of touch with reality. The *rule of reason* approach, although one which can be said to be the most fine-tuned, is too costly and gives little legal certainty. The final approach, that of legality *per se* operating together with a set of screens to weed out pro-competitive tying arrangements, appears the most rational and viable approach. The recommendation is therefore that tying should be seen as *per se* legal. In other words, companies are free to tie until they reach a certain market power threshold. Hereafter the tie will become abusive if it is found to be anti-competitive, and there are no objective justifications which outweigh the anti-competitive effects.

3.   THE FORMAT OF THE APPROACH

This section will discuss whether the current five-step test should remain or whether a new format should be applied. Previous case law is fact-specific, and the Community institutions have no qualms about employing the concept of competition on the merits to their advantage and establishing abuse where they see fit.[24] In effect, currently under EC competition law it is merely sufficient to establish dominance and the form of the abuse, because significant weight has been shifted from the assessment of abuse to the establishment of dominance. Any reform of Article 82, and especially an effect-based reform, should ensure the re-shift of this imbalance to again placing greater emphasis on the abusive conduct and its anti-competitive effects. In comparison, the US approach is better balanced, perceiving the market power requirement as merely *one* of the four conditions for establishing illegal tying.

---

[24]   O'Donoghue and Padilla agree with this position: O'Donoghue and Padilla, *supra* n. 18, p. 177.

Some scholars have argued that an effect-based test should be applied to assess whether conduct is abusive.[25] The consequence of this test would be to commence an abuse analysis with the balancing test of the pro- and anti-competitive effects and remove the initial establishment of dominance, under the assumption that significant competitive harm is proof of dominance.[26] Two initial problems with an effect-based test should be highlighted: first, competitive harm is not necessarily proof of dominance and, second, applying an effect-based test would expand the scope of Article 82, because companies attempting to become dominant will also be caught.[27] These companies' conduct currently falls outside the scope of Article 82.

An effect-based test is clearly leaning more towards a legal *rule of reason* rather than the current moderate illegal *per se* approach. From an economic perspective, the effect-based approach will provide a more realistic picture of the tying. However from a legal perspective, such a test has two major flaws. Firstly, it only works with *ex post* analysis, not *ex ante*, as there needs to be evidence of effects; and, secondly, it eliminates the legal certainty created by a form-based approach with the establishment of dominance. Therefore although it is tempting to suggest that tying cases should commence by assessing the effects of the tying arrangement, it will not solve the issue of legal certainty. Interestingly, none of the economists or lawyers who have dealt with tying in detail has suggested this. In fact, it was made clear in Chapter 2 that market power in tying product markets plays a significant role in assessing tying, and it is an essential element in reviewing whether the tie actually has anti-competitive effects. Moreover, there is one significant hazard associated with commencing with the effect, and that is a very extensive (and time-consuming) analysis of both pro- and anti-competitive effects, which is generally difficult for individual companies to undertake. It is therefore much more sensible to commence a tying analysis by eliminating harmless tying arrangements.

The step-test model applied to tying under EC competition law, US antitrust and patent misuse doctrine offers a much better approach to screen out harmless ties and focus on the anti-competitive ones. The current case law approaches to tying are for clarity summarised in Table 6.1.

The table illustrates that there are some differences between the three approaches; however, they are all relatively small. The need for two separate products test has not played a significant role under the US patent misuse

---

25   Report by EAGCP, 'An Economic Approach to Article 82', July 2005 (http://ec.europa.eu/comm/competition/publications/studies/eagcp_july_21_05.pdf).
26   Monti, G., 'The Concept of Dominance in Article 82' [2006] *European Competition Journal* 312, pp. 44–45.
27   Ibid., p. 45.

Table 6.1   Comparison of current case law approaches to tying

	EC-Competition Law	US Antitrust Law	US-Patent-Misuse-Doctrine
STEP 1	Two separate products (use of consumer demand test and *EC Microsoft* – the demand for the tied product) must have completely ceased	Two separate products (use of consumer demand test and *Eastman Kodak* – demand must be minimised to a point where it is no longer sufficient to sustain competition in either market)	Two separate products (often interpreted literally and along the scope of the patent)
STEP 2	Dominance	Significant market power	Market power (not resulting from the patent alone)
STEP 3	Coercion	Coercion as a result of market power	The licensing or sale of the patented product is made conditional upon purchase or licensing of another (patent or) product
STEP 4	Foreclosure	Substantial amount of commerce on the tied product market is affected	Assessment of 'circumstances'
STEP 5	No objective justifications	No objective justifications	(Staple v non-staple products)

doctrine, and the result is therefore a rather basic assessment and, interestingly, one that does not seem to have taken significant account of consumer demand. The EC competition law approach, on the other hand, is indeed based upon consumer demand, and the CFI in *EC Microsoft* held that it must have ceased completely for the tied product market before the two components can be accepted as one product.[28] This should, though, be contrasted with the Commission's comments in the *Article 82 Guidance*.[29] The threshold under US antitrust is lower. It holds that when the demand for one of the products is so low that offering it separately is no longer sustainable the products are integrated.[30] It acknowledges that some demand will always remain, yet it still requires competition no longer to be efficient in the market before the integrated products can be seen as one.[31]

The second difference between the three approaches is the level of market power, yet case law remains unclear as to whether in fact there is a difference and whether this is truly significant. Importantly, all three approaches have made it clear that a patent (or any IP right) will not in itself be sufficient to establish market power.[32]

The third difference between the US antitrust and the EC competition law approaches to tying is the emphasis which has been placed upon establishing coercion. Under early US case law, coercion was not a separate requirement in the tying analysis; instead, it falls under the two products requirement. In *Times–Picayune*, the Supreme Court stated that an illegal tying arrangement is 'the forced purchase of a second distinct commodity' with the dominant tying product.[33] Therefore, once a second product has been identified, the Supreme Court finds that the condition of coercion is fulfilled. Similarly, under the US misuse doctrine there is a requirement that the licensing or sale of the patented product is conditioned upon purchase or license of another (patent or) product.[34] This condition has not caused any major concerns, as it has often been

---

28   Case T–201/04 *EC Microsoft*, para. 918.
29   'Guidance on the Commission's Enforcement Priorities in Applying Article 82 EC Treaty to Abusive Exclusionary Conduct by Dominant Undertakings', Brussels, 3 December 2008 (http://ec.europa.eu/competition/antitrust/art82/guidance.pdf) (The *Article 82 Guidance*), p. 18.
30   *Eastman Kodak Co. v. Image Technical Services, Inc. et al.*, 504 U.S. 451 (1992) at 462.
31   Ibid., at 462.
32   See *Illinios Tool* at 15–16 and Case 78/70 *Deutsche Grammophon Gesellschaft mbH v Metro-SB-Großmärkte GmbH & Co. KG.* [1971] ECR 487, para. 16.
33   *Times–Picayune Pub. Co. v United States*, 345 U.S. 594, 73 S.Ct. 872, 97 L.Ed. 1277 (1953) at 614.
34   Chapter 5.

part of a contractual agreement between the patentee and the purchaser or licensee. As far as IP rights are concerned, coercion depends on the assessment of whether the patentee has gone beyond the scope of the right, and whether the patentee is applying its patent to shield other products and thereby exclude competitors.

In *Jefferson Parish*, coercion became part of the market power assessment, as the Supreme Court noted that it should be assessed whether consumers were forced to purchase the bundle as a result of the defendant's market power.[35] In *US Microsoft III*, the Court of Appeals included a third element: whether the consumer's choice had been restricted as a result of the company's market power in a *per se* tying violation.[36]

In EC case law, the coercion concept has also revolved around the question whether consumers are left with an economically meaningful choice. In *EC Microsoft*, this was defined as 'the undertaking concerned does not give customers a choice to obtain the tying product without the tied product'.[37] Microsoft's WMP was available free of charge and did not oblige the consumer to use it, and so were other media players, but the Commission still argued that consumers were forced to accept the tied WMP when purchasing Windows.[38] The CFI found these facts irrelevant to establish coercion.[39] The findings were equally good arguments for the technological integration of the WMP with Windows being anti-competitive. This leaves the concept to some extent meaningless, and this conclusion is further reinforced by the fact that the Commission left out the requirement in its *Article 82 Guidance*.[40]

In principle, it is misleading to call *coercion* a requirement for tying; rather, coercion is what tying does – it is the *abuse*.[41] The question is whether customers or consumers are worse off due to the tie: if they are, then it is coercion; if not, the tie is merely a package sale. The concept leads back to the assessment of identifying the type of tying. Hence if the tie is a mixed bundle, the tie is unlikely to be coercion, because the consumer has the possibility of purchasing the bundle as well as the products separately.[42] If, however, the

---

[35] *Jefferson Parish* at 16.

[36] *US Microsoft III* at 85.

[37] *EC Microsoft*, Commission Decision, para. 794 and Case T–201/04 *EC Microsoft*, paras. 944 and 961.

[38] *EC Microsoft*, Commission Decision, para. 831.

[39] Case T–201/04 *EC Microsoft*, para. 970.

[40] The *Article 82 Guidance*, *supra* n. 29, pp. 17–18.

[41] See Faull, Jonathan and Nikpay, Ali (eds), *The EC Law of Competition* (Oxford University Press, Oxford, 2007), p. 167.

[42] There is one reservation, however, and that is that the price of the products individually must not be disproportionate to the extent that it will never make economic sense for the consumer to purchase them individually. If that is the case, the consumer

tying is not mixed bundling, coercion is possible, although an exemption should be made in respect of technological integrated products. Coercion should therefore not be seen as a separate assessment, but is rather the term for the overall finding of an illegal tie. This is also in line with the US patent misuse doctrine, which merely applies the requirement to identify that it is a tying arrangement. The limiting of consumer choice is an important indicator of abuse, but should not on its own constitute an antitrust concern.[43] Therefore, it would be better placed to assess consumer choice under the anti-competitive effects condition.

All three approaches acknowledge implicit or explicit that there is a need to demonstrate (future/likely) harm. As will be discussed, the standard of proof is low, and there is no clear guidance as to how much harm or foreclosure it is necessary to prove, and whether this needs to be actual, potential, or likely. Chapter 2 identified the importance of this requirement and therefore, despite the vagueness surrounding this step in the current approaches, it is clear that it should play a leading role in the new suggested model.

The final step in the current approaches is that of no objective justifications. This is perhaps the only step in the tests where US antitrust and EC competition law are completely aligned, as both have very high thresholds which appear impossible to reach. In respect of format, it is essential to maintain an assessment of potential objective justifications if a more lenient approach to tying is to make sense. It plays an important role alongside the anti-competitive effects test and should, in fact, be given more weight than currently is the case under US antitrust and EC competition law, especially, if tying arrangements should no longer be seen as *per se* illegal.

3.1. Recommendation: a Four-step Test

The above discussion, together with the findings in section 2 that tying should be considered legal *per se* and certain screens should be introduced to weed out pro-competitive tying arrangements, leads to the recommendation that the step format of the current approach to tying should remain, because it is ideal to achieve an approach which can incorporate the relevant measures identified to ensure a more flexible and positive approach to tying in line with economic thinking. The following four step test to tying is therefore suggested:

---

is indirectly forced to take the bundle. See Commission Decision *Digital Undertaking*, (1997) *XXVIIth Report on Competition Policy*.

[43]   Monti, G., *EC Competition Law* (Cambridge University Press, Cambridge, 2007), p. 191.

(1)   Are there two separate products?
(2)   Does the company hold significant market power in the tying product market?
(3)   Are there any anti-competitive effects?
(4)   Are there no objective justifications?

As can be observed, the test is very similar to both the current US and EC approaches, albeit that it excludes the need for assessing coercion and in the process becomes a more simple test. What becomes clear is that it is in fact not the format of the approach which creates a faulty result for tying arrangements, but rather the individual steps and the test within that are in need of fine-tuning. For instance, it is clear that too many tying arrangements are not cleared at the first two screens, those of whether there are two separate products and whether the company holds significant market power. These individual steps need to be tightened up to provide a more meaningful approach to tying.

4.   STEP 1: THE NEED FOR TWO SEPARATE PRODUCTS – ALTERNATIVE SOLUTIONS

4.1.   Introduction

This section will assess alternative options to the current test of two separate products in tying cases, which have been based on whether there is a separate consumer demand and independent manufactures of the tied product. The current test is too simplistic – it does not offer a realistic view of the markets or the products, when these are more complex and innovative. Moreover, it is closely linked to the establishment of dominance, and as such does not seem to offer a fair picture of the relevant product market, but rather is pre-set to identify certain features. One of these is the scope of the IP rights regardless of any evidence of different product definitions.

As was illustrated by Bork, there must be a point where it becomes absurd to divide the product into smaller parts because it is unacceptable not only for companies but also for the consumer.[44] The competition authorities have relied heavily on the argument that tying will restrict consumer choice.[45]

---

[44]   Bork, Robert H., *The Antitrust Paradox, A Policy at War with Itself* (Basic Books, the Free Press, New York, 1993), pp. 378–379.
[45]   *EC Microsoft*, Commission Decision, paras. 826–834 and *Jefferson Parish* at 12–14.

However, if a purchase becomes too complicated due to too many choices, this may deter consumers from making the purchase.[46] Therefore, there is a need to strike a balance between sufficient consumer choice and the need to allow for complex products to remain as one unit if the competition authorities are to regulate tying arrangements efficiently.

The courts on both sides of the Atlantic have offered an alternative test to assess whether a product is just one product or a tying arrangement. The most basic approach is the literary approach applied mainly in early US case law which makes sense in contractual tying cases where products are clearly defined such as salt and salt processing machines, railways and land, and sugar and the delivery of sugar. Where courts have been forced to assess more complex products and market structures they introduced a consumer demand test for assessing whether there was a separate tied product.[47] Relying on consumer perception of the products the court could establish whether a product was in fact two. The US courts have looked at whether there were independent manufacturers in the tied product market as an indicator of whether consumers perceive the products as separate. In other words, if there is sufficient demand so that it is efficient for the company to provide the tied product, there is a tying arrangement.[48]

In comparison, in *EC Microsoft* the CFI set an extremely high standard of proof for when a product can be seen as a component of another product. It stated that this is the case where there is 'absence of independent demand'.[49] In other words, as long as there is a demand for the component on its own, it will never be accepted by the EC Courts to have been integrated, but merely bundled. This is a higher threshold than the one permitted under US antitrust law. It is also inflexible, as it does not take into account product development and potential aftermarkets. In contrast, the Commission in its *Article 82 Guidance* appears to have moved to a more realistic threshold in line with the US approach.[50] Yet uncertainty remains as to how much consumer demand is necessary under EC competition law. The competition authorities have also at times relied on the companies' own perception of the products[51] and this has

---

46   Nalebuff, Barry, 'Bundling, Tying, and Portfolio Effects', DTI Economics Paper No 1, Part 1 – Conceptual Issues, February 2003, p. 32; and Dhar, R. and Nowles, S.M., 'To Buy or Not to Buy: Response Mode Effects on Consumer Choice' (2004) 41 *Journal of Marketing Research* 423, pp. 431–432.
47   *Jefferson Parish* at 21–22 and Case T–30/89 *Hilti*, para. 67.
48   *Eastman Kodak*, at 462.
49   Case T–201/04 *EC Microsoft*, para. 918.
50   The *Article 82 Guidelines*, *supra* n. 29, p. 18.
51   *EC Microsoft* Commission Decision, paras. 805 and 813; *Jefferson Parish* at 21–22; *Eastman Kodak* at 457–459.

stretched to the consideration of market behaviour and conditions.[52] Interestingly, in the EC these facts have always been applied to the defendant's detriment, whereas in the US they have played a crucial role in establishing pro-competitive effects of the tying arrangement.

For technological integration two approaches have been applied by the courts. The first is based upon establishing whether the integrated products created pro-competitive effects and technological efficiencies, 'some advantages' or 'superior technical product performance', which could not be gained from the mere bolting of the two products together.[53] However, the US courts have been reluctant fully to engage in a technical assessment of the products, and have instead assumed that the defendant has met this standard.[54] In *EC Microsoft* it was Microsoft's own inability to provide any acceptable efficiencies that restricted a fully-fledged assessment from occurring.[55] Therefore under both jurisdictions the question of the standard of proof regarding technological integration has been left unanswered.

The other approach is simply to apply a *rule of reason* assessment, which in effect means that a thorough discussion of whether the products are integrated or not will be replaced with an assessment of the pro- and anti-competitive effects created by the tying arrangement.[56] On balance the anti-competitive effects must outweigh the benefits generated from the tying arrangement, demonstrating that the conduct unreasonably restrains competition.[57]

The case law provides two important issues of the separate products test which must be taken into consideration in a new test; first, there are three perspectives, the consumers, the manufacturers (and within this the market conditions) and finally an objective assessment of the actual tying arrangement, all of which need to play a role within a new separate product test. Second, case law indicates the possibility of a distinction in the assessment of contractual tying and technological integration.

A particular concern, which appeared in *EC Microsoft* was the early conclusion by the Commission of the presence of two separate product. This permitted the Commission subsequently to dismiss Microsoft's attempt to defend the tying arrangement as technological integration, although it had never truly

---

52   *Jerrold Electronics* at 555–556 and Case C–333/94 *Tetra Pak II*, para. 36.
53   *US Microsoft II* at 949–952 and Case T–201/04 *EC Microsoft*, para. 1159.
54   *US Microsoft II* at 949–950 and *ILC Peripherals Leasing Corp. v International Business Machines Corp.* 458 F. Supp. 223, 439 (N.D. Cal. 1978) (*ILC Peripherals Leasing II*).
55   Case T–201/04 *EC Microsoft*, para. 1160.
56   *US Microsoft III* at 95–96.
57   Ibid.

given this any consideration under the separate product test. While the CFI supported the Commission's conclusions that the operating system and media player were separate products and ignored Microsoft's valid arguments regarding the lack of consumer demand for the operating system without a media player,[58] it did provide Microsoft with an opportunity to defend its tying arrangement if Microsoft could show that its technological integration offered 'superior technology' to that of the mere tying. However, it is clear that to ensure a realistic opportunity for the defendant to defend the tying arrangement the defendant must be permitted to rely on a technological integration argument. This should be done as part of the separate product test rather than as part of the objective justification assessment. Otherwise, as illustrated by *EC Microsoft* the competition authorities will be able to dismiss such a claim based merely on its own narrow market definition. The alternative solutions given below are based on these conclusions and the ones reached in Chapter 2. These highlighted that not only was there a need to create a safe haven for certain tying arrangements, but this needed to be based on market power held by the tying company and on the understanding that innovative and technologically complex products required different treatment from traditional products. Thus, a test which establishes when components of a product are fully integrated into that product needs to be identified. The starting point for alternative solutions is to question whether we in fact need a separate products test. Is such a test an absolute for tying cases or can it be done away with?

The second analysis is of the new Technology Transfer Block Exemption Regulation[59] (the TTBER) and the US IP Licensing Guidelines,[60] as they offer a different view from the market definition when IP rights are involved by taking into consideration that products can consist of components.

The third alternative is the indispensability test, which was developed in *Oscar Bronner*.[61] It offers an alternative to the consumer demand test because it perceives the products from a competitor's perspective rather than that of the consumer. Finally, two suggestions are made based on an analysis of technological integration and when two products become one: the broken product test and the expanded consumer demand test.

---

58 Case T–201/04 *EC Microsoft*, paras. 913, 921–923, and 944.

59 Commission Regulation (EC) No 772/2004 of 27 April 2004 on the application of Article 81(3) of the Treaty to categories of technology transfer agreements [2004] OJ L123/11.

60 Antitrust Guidelines for the Licensing of Intellectual Property, issued by the US Department of Justice and the Federal Trade Commission, 5 April 1995 (http://www.usdoj.gov/atr/public/guidelines/0558.htm).

61 Case C–7/97 *Oscar Bronner GmbH & Co. KG v Mediaprint* [1998] ECR I–7791, [1999] 4 CMLR 112.

4.2.　Is the Separate Products Test Essential?

The inclusion of the separate products test is in effect a result of the form-based approach applied to exclusionary conduct. The separate product test is used only in tying cases and therefore, if tying moved to a *rule of reason* assessment under US antitrust law and a radical reform of Article 82 cases were to be adopted, where the main goal is to establish purely a more economic approach it would be a logic consequence to eradicate the separate product test. There are in fact some arguments which could be made to favour such abolition. For example, Section 2 of the Sherman Act, which deals with monopolisation and attempt to monopolise, also catches tying arrangements but in effect does so under different conditions, which therefore avoids an assessment of the product and the question whether there are two products. Instead, it looks at the monopolist's behaviour to assess whether it unreasonably preserved its monopoly position or enhanced its market power in case of attempt to monopolise. In its *Section 2 Report* the DoJ does not include a separate product test in its final recommendations for tying.[62] There is therefore in principle no reason why a similar approach could not be adopted for Article 82 and Section 1 tying arrangements for that matter. However, the market power requirement for monopolisation under Section 2 has a much higher threshold, one which would permit many tying arrangements that are now caught under Article 82 and Section 1.

The approach in Section 2 is very similar to the sacrifice test, which shall be discussed further below, and merely focuses on the anti-competitive effects of the behaviour, while at the same time assuming that the purpose of the tying arrangement is exclusionary.[63] The test does therefore not give sufficient consideration to product developments and other efficiencies which only tying can achieve.[64] Moreover, Section 2 requires a thorough assessment of the facts of each case; in comparison, the separate product test under Article 82 and Section 1 works as a first step in a screening procedure to weed out pro-competitive cases. This would also be its purpose in the proposed new approach to tying. It creates a safe harbour and some level of legal certainty. Without a separate product test, tying arrangements would have to be assessed on a case-by-case basis, an approach notorious for being time consuming and expensive. Interestingly, none of the economists whose theories and thoughts were examined in Chapter 2 advocated the removal of the separate product test, although advocating a more lenient approach to tying arrangements in general.

---

[62]　*Section 2 Report, supra* n. 1, p. 90.
[63]　See Werden, Gregory J., 'Identifying Exclusionary Conduct under Section 2: The "No Economic Sense" Test' (2006) 73 *Antitrust Law Journal* 413.
[64]　Ibid., p. 414.

The separate products test could be eradicated if a one-rule-fits-all were to be adopted for exclusionary behaviour in the future;[65] however, even such a rule will require special exceptions for particular conduct such as the introduction of new products,[66] and therefore it makes sense, and not least permits some legal certainty, to contain the separate product test, albeit with a more economic approach to take into account innovation and product development. The main reasons for the separate product test in the first place appears old-fashioned and wooden.

4.3. The TTBER and the US IP Licensing Guidelines – Alternative Market Definitions?

Because so many tying cases include some form of IP rights, it makes sense to look towards a form of competition regulation which successfully deals with IP rights. The US IP Licensing Guidelines and the new TTBER under Article 81 do exactly that and take a more realistic approach to the market definition when dealing with IP rights. They require not only the identification of the relevant product market but also the relevant technology market and the US IP Licensing Guidelines have also introduced an innovation market.[67] They both deal effectively with the transfer of technology; that is the process of technology IP licensing.

In respect of the TTBER, the TTA Guidelines clarify that a product can contain a technology input, and it may be this that holds the IP right rather than the whole product as such. It is noteworthy that the TTBER, unlike the Commission Notice on the definition of the relevant market, defines a product. The definition is not particularly helpful for our purpose, except for the fact that it includes some components of products, here referred to as intermediary goods: '"product" means a good or a service, including both intermediary goods and services and final goods and services'.[68] This definition corresponds well with Dolmans and Graf's idea of focusing on the correct level of the market discussed in Chapter 3.[69]

---

[65]   The *Article 82 Guidance, supra*, n. 29, gives no such indication as it contains an assessment of specific types of abuse.

[66]   Werden, *supra* n. 63, p. 414.

[67]   See TTBER, *supra* n. 59, Article 1(1)(j) and Article 3 and Commission Notice, Guidelines on the application of Article 81 of the EC Treaty to technology transfer agreements [2004] OJ C101/02, paras. 19–22 (the TTA Guidelines) and US IP Licensing Guidelines, *supra* n. 60, Section 3.2.3.

[68]   TTBER, *supra* n. 59, Article 1(1)(e).

[69]   Dolmans, M. and Graf, T., 'Analysis of Tying Under Article 82 EC: The European Commission's Microsoft Decision in Perspective' (2004) 27 *World Competition* 225, p. 228.

The TTBER does not define a technology. However, it is defined in the Guidelines on Horizontal Cooperation Agreements as 'the [IP] that is licensed and its close substitutes',[70] which proves the point argued above that the Commission sees the scope of the IP right very much as a substitute for a realistic market definition assessment. Anderman and Kallaugher suggest that the technology should instead be defined around functionality from the perspective of the potential licensee.[71] They argue that this would take the emphasis away from IP rights, although there are also downsides to this approach as it will require a more detailed technology assessment in some cases.[72]

In the TTA Guidelines we are treated to a definition of the technology market as 'other technologies which are regarded by the licensees as interchangeable with or substitutable for the licensed technology, by reason of the technologies' characteristics, their royalties and their intended use'.[73] This test is essentially the market definition test also applied to ordinary products; however, the TTA Guidelines allow another method of defining the technology market by looking at the market for the products which incorporate the licensed technology.[74] The Commission will look at the technology in relation to the product market, but also at whether there are any substitute technologies to assess the competitive restraints on the technology.[75] In other words, the focus is upon the manufacturers/the supply side. Anderman and Kallaugher note that this division is merely to differentiate between the market for the sale/licence of the technology and the market for the sale of products, either produced with the licensed technology or protected by IP rights.[76] If this is the case, the approach in the TTBER and TTA Guidelines appears to be remarkably different from the approach taken by the Guidelines on Horizontal Cooperation Agreements. This is unfortunate, as the Commission then appears to have two sets of policies for the same issues, depending on the type of conduct the IP owners are involved in.

In discussing substitutable IP rights, Anderman and Kallaugher suggests that:

---

[70] Commission Notice Guidelines on the applicability of Article 81 of the EC Treaty to horizontal cooperation Agreements (2001/C 3/02), January 2001, para. 47.

[71] Anderman, S. and Kallaugher, J., *Technology Transfer and the New EU Competition Rules, Intellectual Property Licensing after Modernisation* (Oxford University Press, Oxford, 2006), s. 6.59.

[72] Anderman and Kallaugher, *supra* n. 71, s. 6.59.

[73] The TTA Guidelines, *supra* n 67, para. 22.

[74] Ibid.

[75] Ibid., para. 23.

[76] Anderman and Kallaugher, *supra* n. 71, s. 6.40.

[where] a patent is an essential element of a technology, the owner of that patent will have market power, but the market in which that power is exercised is a market for the technology (and any competing technology) not the 'market' for the bit of technology covered by the essential patent.[77]

It is hoped that this is the way the Commission wishes its TTA Guidelines to be interpreted.

The TTBER only recognises actual competitors as 'competitors in the technology market',[78] whereas in the product market both actual and potential competitors need to be taken into account. This restriction shows that it is difficult to identify competitors, and even more so potential competitors, that is those who are likely to invent a component or already have interchangeable technology, which is not in use or licensed out, which can compete with the current technology.[79] From a dominance perspective, the IP holder will be made worse off by the fact that potential competitors on the technology market are not taken into account. Moreover, it may be unrealistic to expect patent owners to identify competitors owing to the nature of the patent, whereas it is more likely that copyright holders will be able to do so. For instance, Hilti would have had difficulties in doing so, as its patent on the cartridge strips protected it from direct competition, although under a TTBER approach the Commission should have assessed whether substitute technologies were applied in substitute products, that is other PAFS.

In contrast, the Commission identified several competitors when assessing the media player market in *EC Microsoft*,[80] although it accepted only 'media players with similar functionalities as direct competitors',[81] and thus excluded others from the market.[82] On the other hand, the IP owner may benefit from the more specific market definition which deliberately distinguishes between the technology component of the product and the actual product, if it diminishes her market shares on the product market.

---

[77] Ibid., s. 6.57

[78] TTBER, *supra* n. 59, Article 1(1)(j): 'competing undertakings on the relevant technology market, being undertakings which license out competing technologies without infringing each others' intellectual property rights.' However, compare with TTA Guidelines, *supra* n. 67, para. 30 defining potential competitors in the technology market: '[t]he parties are considered to be potential competitors on the technology market where they own substitutable technologies if in the specific case the licensee is not licensing his own technology, provided that he would be likely to do so in the event of a small but permanent increase in technology prices'.

[79] Ibid.

[80] *EC Microsoft* Commission Decision, paras. 411–415.

[81] Ibid.

[82] Ibid., paras. 141–143.

It is important to emphasise that the approach in the TTBER is no more lenient towards IP holders. However, there is no doubt that it offers a more realistic (and flexible) product/technology market definition when dealing with IP rights, as it takes into consideration the fact that a technology can be a component of a product and it assesses this without focusing on the IP protection of the technology which would normally shape the relevant market definition. Moreover, a patent must cover the essential components of a product in order to be a significant barrier to entry and influence supply elasticity.[83] The approach also acknowledges that the technology may be applied in various products, which may or may not be part of the same market. For instance, in the media player market, iTunes, which is a software program linked to the web that downloads music to iPods (a form of MP3 player), has similar technology to that applied by the standard media players, such as Windows Media Player and RealPlayer. It would therefore be a competitor on the technology market with these other media players, but is not considered as such in the product market, and was not included in the assessment in *EC Microsoft*.[84] Other media players were equally excluded from the product market due to the fact that they were based on different technologies and were 'business to consumer'-oriented rather than 'business to business', like Microsoft.[85] Only the latter point could risk excluding them from the product market following the TTBER market definition approach.

Nevertheless, identifying the product market in high-technology industries is difficult as these are characterised by rapid innovation – hence expecting companies to be able to identify a technology market as well may cause further complications.[86] In addition, although the TTBER accepts that in principle there can be competing technologies,[87] these are more likely to be found among technologies protected by other forms of IP rights than patents, making it almost certain that patent owners will be found dominant or even as having a *de facto* monopoly in the technology market. Thus, the TTBER approach does not deviate from the general assumption that patents, in particular, may create automatic market power as found in case law.

---

[83]   Rahnasto, Ilkka, *Intellectual Property Rights, External Effects and Antitrust Law, Levering IPRs in the Communications Industry* (Oxford University Press, Oxford, 2003), p. 27.

[84]   Although mentioned in para. 138 in *EC Microsoft* Commission Decision, as part of Apple as competitor, it is not mentioned further.

[85]   Ibid., para. 143.

[86]   Dolmans, Maurits and Piilola, Anu, 'The Proposed New Technology Transfer Block Exemption, Is Europe really better off than with the current regulation?' (2003) 26 *World Competition* 541, p. 553.

[87]   TTA Guidelines, *supra* n. 67, para. 24.

Interestingly, in refusal to supply cases, the ECJ referred to the upstream product as a raw material – an input needed to produce the downstream product.[88] The idea of the upstream product forming only part, albeit an essential part, of the downstream product is not developed further by the ECJ in those cases as regards the relevant market/product definition scenario. However, the similarity with the TTBER's technology market is remarkable. Thus, some aspects of the TTBER approach could easily be transferred to or anticipated in Article 82 cases.[89] It is therefore possible that the Commission could be persuaded to apply an approach similar to that under Article 81 to future tying cases under Article 82.[90]

Revisiting case law such as *Hilti* with the TTBER form of market definition, the product market would be the PAFS, whereas the technology market would be the cartridge strips market and the nail gun market, where Hilti held patents. Hilti would most likely still have been found to have market power in the technology market for the cartridge strips, but not necessarily in the product market, as notwithstanding the IP rights (both patents and copyrights) Hilti had on the cartridge strips there were nevertheless two other producers of PAFS cartridge strips in the Community alone.[91] Despite AG Jacobs' contention, neither the Commission nor the EC Courts ever assessed this properly.[92] It would follow from this that if AG Jacobs' comment in *Hilti* was penetrating, it would be a question of substitutable technologies applied for PAFS and substitutable PAFS, of which there were a few,[93] rather than the particular patent, which would be the determining factor in finding whether Hilti was dominant or not. In *Tetra Pak II*, however, the dominance held on both aseptic markets (the aseptic machine market constituting the relevant technology market) was most likely to be too great to change the outcome of the case when applying the TTBER approach. In *Hilti* and *Tetra Pak II* the products and technologies were still relatively too basic to be able to apply the TTBER approach without too much complication. In *EC Microsoft*, however, applying

---

[88] Cases C–241–2/91P *Radio Telefis Eireann v Commission* (*Magill*) [1995] ECR I–743, [1995] 4 CMLR 718 and C–7/97 *Oscar Bronner*.

[89] See Case C–418/01 *IMS Health GmbH & Co. OHG v NDC Health GmbH & Co. KG* [2004] ECR I–5039 [2004] 4 CMLR 1543.

[90] See Kroes, Neelie, 'Preliminary Thoughts on Policy Review of Article 82', Speech at the Fordham Corporate Law Institute, New York, 23 September 2005 (http://europa.eu/rapid/pressReleasesAction.do?reference=SPEECH/05/537&format=HTML&aged=0&language=EN&guiLanguage=en).

[91] Case T–30/89 *Hilti v Commission* [1991] ECR II–1439, [1992] 4 CMLR 16, paras 52–53.

[92] AG Jacobs' Opinion in Case C–53/92P *Hilti AG v Commission* [1994] ECR I–667, [1994] 4 CMLR 614, p. 624.

[93] Case T–30/89 *Hilti*, paras 52–53.

the TTBER, the operating system would be the technology tying the media player, the product, to it. This is a rather artificial assumption as, within both the operating system and the media player, there are in fact technologies which are seen as components of these products, but the test does nothing to answer whether the media player has been truly integrated with the operating system. This is because the TTBER does not clarify the meaning of either technology or product. However, it has been suggested that a barrier to entry to the product market is most likely to occur if the (IP-protected) component is essential to the commercial product.[94] This leaves us with two interesting conclusions: firstly, WMP is not, according to the Commission, essential to the functioning of Windows,[95] whereas Microsoft in fact argued the opposite, namely that if the WMP was removed from Windows, essential functions of the operating system would no longer work.[96] Secondly, returning to Anderman and Kallaugher's argument above that the technology should be defined on the basis of the functionality from the perspective of the potential licensees means that certain products like Windows with WMP may require a bundle of technologies to constitute a product.[97] In other words, even though we can identify several technologies within one product, that does not necessarily mean that there are several products.

Although the TTBER offers an alternative, more refined and realistic market definition, in particular when dealing with IP rights and technologically complex products, it does not offer a solution to identify when a component is just part of a product and when it is a stand-alone product. In comparison, the US IP Licensing Guidelines apply, as mentioned, two additional dimensions to the relevant market definition in respect of IP rights: the technology market – that is the market of licensed IP – and an innovation market – that is research and development towards new or improved goods or processes.[98] The technology market is applied where the IP is marketed separately from the products in which it is used and it allows the DoJ and the FTC better to assess the competitive effects of a licensing agreement.[99] Where this is not the case, the US IP Licensing Guidelines note that it is unnecessary to

---

[94]   Rahnasto, *supra* n. 83, p. 27.

[95]   *EC Microsoft* Commission Decision, para. 809.

[96]   Ibid., para. 1019.

[97]   Anderman and Kallaugher, *supra* n. 71, s. 6.59.

[98]   US IP Licensing Guidelines, *supra* n. 60, Section 3.2.3. 'Innovation market analysis' has also been applied to mergers: see Horizontal Merger Guidelines, issued by Department of Justice and Federal Trade Commission, 2 April 1992, Revised: 8 April 1997, para. 1.521, and Glader, M., *Innovation Markets and Competition Analysis, EU Competition Law and US Antitrust Law* (Edward Elgar, Cheltenham, 2006), p. 72.

[99]   US IP Licensing Guidelines, *supra* n. 60, Section 3.2.2.

make a separate analysis of the technology market.[100] Most tying cases would thus fall within the latter category. However, the product could be broader in scope than the IP-protected technology. It appears that even if the product is broader in scope, the DoJ and the FTC will not attempt to split up the product and claim that there is a tying arrangement between the IP protected technology and the other unprotected components of that product.

In situations where a licensing agreement negatively affects competition in the development of new or improved products, the US IP Licensing Guidelines propose to identify an innovation market. This would be the case where the licensing agreement may affect the development of products that do not yet exist or in geographic markets where there is no competition or it is likely to develop.[101] This is a refined way of taking into account potential competitors of a technology as these can be extremely difficult to identify from either a product-market or even a technology-market perspective. The US IP Licensing Guidelines, with the inclusion of an innovation market, anticipated some of the main concerns that were raised in *US Microsoft III*,[102] such as the question whether a technologically integrated product enhances consumer welfare or restricts innovation in the tied market. Antitrust case law has thus treated technological integration more favourably than contractual tying by the courts.[103] This distinction between contractual tying arrangements and technological integration made in case law seems to be at odds with the US IP Licensing Guidelines. According to the US IP Licensing Guidelines, if the technological component of the product is in some way found to affect future innovation, in particular by not being available as a separate product, the licensing arrangement will be scrutinised by the antitrust agencies for anti-competitive effects and may be challenged.[104]

The inclusion of both the technology market and the innovation market in the assessment of licensing of technology is an interesting development of the relevant market definition, offering a more thorough assessment as well as perhaps a more realistic assessment of what is a market and product. However, just as was the case with the TTBER there is no doubt that these additional markets create a more complex picture, and this is likely further to complicate companies' own abilities to assess their position in an antitrust environment. Both must therefore be dismissed as alternative solutions to the two separate product test.

---

[100]   Ibid., n. 19.
[101]   Ibid., Section 3.2.3.
[102]   *US Microsoft III* at 94–95.
[103]   Ibid., at 84 and 95.
[104]   US IP Licensing Guidelines, *supra* n. 57, Sections 3.2.3 and 5.3.

4.4.   The Indispensability Test – an Alternative to the Consumer Demand Test?

We need to turn away from IP rights and look at similar non-reproductive product cases to find a more flexible attitude by the Commission and the EC Courts. For instance, *European Night Services*,[105] which actually arose under Article 81, revolved round a joint venture providing overnight passenger rail services between the UK and the rest of Europe via the Channel Tunnel. The CFI held that a facility could not be classified as essential if there were substitutes; having a mere competitive advantage over others was not sufficient.[106] *Oscar Bronner* offers an alternative solution. In *Oscar Bronner*, a small newspaper wanted to make use of a bigger newspaper's delivery system, but the latter refused to grant this access. The case is interesting and relevant because it relied heavily on *Magill*, which dealt with the interface between competition law and IP rights – more specifically with access to some copyright-protected TV listings to make a TV guide listing of three broadcasters' programmes. The ECJ had found that in exceptional circumstances a refusal to license IP rights could be abusive. These exceptional circumstances were present in *Magill* as the broadcaster held a *de facto* monopoly on the TV listings market, which it used to exclude competition in a secondary market by denying access to an indispensable raw material. The refusal, for which there was no objective justification, hindered the appearance of a new product for which there was a consumer demand.[107]

Oscar Bronner further developed this exceptional circumstances test. It focused on the issue of indispensability, holding that a product is indispensable inasmuch as there is no actual or potential substitute for it and the refusal would eliminate all competition on the part of the person requesting access.[108] Substitutes would include less advantageous products.[109] The ECJ continued to note that the requirement of indispensability can only be satisfied if there are technical, legal or economic obstacles capable of making it impossible, or even unreasonably difficult, for another to produce the product.[110] Finally,

---

[105]   Cases T–374–375, 384 and 388/94 *European Night Services v Commission* [1998] ECR II–3141, [1998] 5 CMLR 718; the judgment was delivered after the opinion of AG Jacobs in Case C–7/97 *Oscar Bronner GmbH & Co. KG v Mediaprint* [1998] ECR I–7791, [1999] 4 CMLR 112.

[106]   *European Night Services*, paras. 215–217 and 221.

[107]   Case C–241–242/91P *Magill*, paras 53–57.

[108]   Case C–7/97 *Oscar Bronner*, para. 41; see also Monti, G., 'Article 82 EC and New Economy Markets', Chapter 2 in Graham, C. and Smith, F. (eds), *Competition, Regulation and the New Economy* (Hart Publishing, Oxford, 2004), pp. 43–44.

[109]   Case C–7/97 *Oscar Bronner*, para. 43.

[110]   Ibid., para. 44.

although contradicting itself somewhat, the ECJ held that upon assessment of indispensability it was not enough to argue that it was not economically viable for a small undertaking to produce the product.[111] 'For such access to be capable of being regarded as indispensable, it would be necessary at the very least to establish . . . that it is not economically viable to create a second [product or service] with a [sale] comparable to that of . . . the existing [product or service].'[112] The ECJ's ruling can be interpreted as holding that the elements of the test for indispensability do not need to be applied cumulatively, but can be applied individually to find indispensability,[113] save when the obstacle is economic in character. In respect of economic inability, it has to be measured in respect of the product or service being on a similar size as that of the dominant company. As reworded, the test would be: the requirement of indispensability is fulfilled if there are technical or legal obstacles making it impossible or unreasonably difficult to produce the product.[114] Upon assessing this, one can take economic viability into consideration; however, this cannot alone constitute a valid reason, unless it is not economically viable to create a second product or service of a similar size.[115]

The question is: could this test for indispensability be transferred to tying cases to the effect that tying would only be truly illegal if the tying product was *indispensable*, that is if there were no substitutes? Importantly, Drexl points out that it is necessary to distinguish between indispensable and *successful* technology.[116] Some technology may appear indispensable due to the fact that it is successful and thus popular among customers to the extent that they prefer that technology; however, that will not make the technology irreplaceable. In accepting this argument, the worry is that a competition authority may not be able to distinguish between these two, simply through lack of expertise in the technology, and will therefore interfere unnecessarily in the market, restricting the actions of an efficient company and its popular product.[117] This concern has been recognised in US case law.[118] Furthermore,

---

[111]   Ibid., para. 45.
[112]   Ibid., para. 46.
[113]   The 'or' in Case C–7/97 *Oscar Bronner*, para. 44 indicates that this is the correct interpretation.
[114]   Ibid., para. 46.
[115]   Ibid., paras. 45–46.
[116]   Drexl, Josef, 'Refusal to License and IP Right', – Comments by the Max Planck Institute for Intellectual Property and Tax Law, Munich (http://ec.europa.eu/comm/competition/antitrust/others/drexl.pdf), slide 6.
[117]   Harchuck, K. E., 'Microsoft IV: The Dangers to Innovation posed by the Irresponsible Application of a Rule of Reason Analysis to Product Design Claims' (2002) 97 *Northwestern University Law Review* 395, pp. 396 and 431.
[118]   *ILC Peripherals Leasing II.*

not acknowledging this distinction would mean that successful companies would risk interference, merely due to their success. Turning upon such companies would remove incentives to be both competitive and innovative.[119]

Applying the test of indispensability to tying is justified by the fact that it originated under the refusal to supply/licensing cases, which are closely linked to tying cases. Some refusals were based upon pressuring customers also to purchase or license a secondary product or service.[120] Moreover, the test is similar to the approach of the US patent misuse doctrine, because if the tying product is indispensable, as it will be if it is patented, then one may not add the tied product to it, as that would harm consumer access to the tied product. Equally, the patent misuse doctrine operates under the same philosophy that the patent owner may not attempt to extend his exclusive rights beyond the patent by tying a product to the patented product.[121] Although on the surface this test appears to solve the problem of when a component is a separate product and not integrated, this would only be the case where the component was truly indispensable for the tied product. The test focuses merely on the tying product, and does therefore not say anything about whether the tied product is equally indispensable for the tying product. It also does not answer clearly whether there are in fact two products. Consequently, what can only be concluded is that if the tying product is indispensable, the company in question is most likely to hold a monopoly or a near-monopoly position on the tying product market, and therefore this test is probably more appropriate for the establishment of dominance. The test must therefore be rejected.

4.5. The Expanded Consumer Demand Test and the Broken Product Test

As illustrated, the current consumer demand test is not able to distinguish between a product and components of that product. The consumer demand test will continue to divide the product into smaller product units until theoretically there is no consumer demand for the components. This approach is naturally unrealistic as there is in principle always a demand for spare parts or

---

[119]   'The successful competitor, having been urged to compete, must not be turned upon when he wins': *United States v Aluminum Co. of America (Alcoa)*, 148 F 2d 416 (2nd Cir. 1945) at 430.

[120]   See *Napier Brown/British Sugar*, Commission Decision 88/519/ EEC [1988] OJ L284 and Case 311/84 *Centre Belge d'Etudes du Marché-Télémarketing v Compagnie Luxembourgeoise de Télédiffusion SA and Information Publicité Benelux SA* [1985] ECR 3261, [1986] 2 CMLR 558.

[121]   *Virginia Panel Corp. v MAC Panel Corp.*, 133 F.3d 860 (Fed. Cir. 1997) at 869.

individual components of a product, if these parts of the product break. However, that does not necessarily make them a product; there may be merely a demand for them as spare parts.[122] The Commission seems to apply the market definition as a means of regulating companies and competition rather than a realistic assessment of the market.[123] As was illustrated in *Soda-ash*, where the Commission clarified that it would identify the relevant market and a company's market power by looking at the 'area of business' of the alleged dominant undertaking and identify existing and potential competitors, but also take into account 'the nature of the abuse which is being alleged and of the particular manner in which competition is impaired in the case in question'.[124] By assessing the nature of the abuse, the Commission does not apply a strictly objective assessment of the relevant market, but instead a strategic market definition.

Rahnasto quite critically notes in respect of IP protected products that there is a perception that IP protected products have market power. This has practical implications for the relevant market definition and has resulted in a legal presumption that these products hold dominant positions for the purpose of Article 82 infringements.[125] There is therefore a conflict between the view of the regulators, the industries, and consumers, which is worth investigating. 'From a commercial point of view, firms consider decisions such as the selection of product and product packaging to be the essence of their commercial judgement in marketing generally and in the exploitation of intellectual property rights in particular.'[126] On the other hand, the Commission and the EC Courts take another view: that the choice of product market, that is whether established as a product package or individual components, is subject to regulation under the competition rules.[127]

The very essence of this problem was outlined in the *Hilti* case.[128] It highlights that there can be a difference in how companies and consumers perceive or define a product in comparison to how competition authorities understand or define products for the purpose of regulating competition, despite the fact that the competition authorities attempt to distinguish products based on a

---

[122]   Dolmans and Graf, *supra* n. 69, p. 228 and Evans, D.S. and Padilla, A.J., 'Tying Under Article 82EC and the Microsoft Decision: A Comment of Dolmans and Graf' [2005] *World Competition: Law and Economics Review*, p. 9.

[123]   Anderman, Steve, *EC Competition Law and Intellectual Property Rights, the Regulation of Innovation* (Oxford University Press, Oxford, 1998), p. 161.

[124]   Commission Decision, *Soda-ash – Solvay* [1991] OJ L152/21, para. 42

[125]   Rahnasto, *supra* n. 83, p. 23.

[126]   Anderman, *supra* n. 123, p. 154.

[127]   *Hilti* Commission Decision; Case C–53/92P *Hilti AG v Commission* [1994] ECR I–667, [1994] 4 CMLR 614, and Anderman, *supra* n. 123, p. 154.

[128]   Cases T–30/89 *Hilti* and C–53/92P *Hilti*.

consumer perspective. In *Hilti*, Hilti saw its PAFS as a product, whereas the Commission and the EC Courts did not. The definition of a product will be given when dealing with basic or normal products such as bananas, bicycles and books; on the other hand, when the product becomes more complex or part of the most recent technology, the definition of a product becomes essential in order to understand whether there are two products or components comprising one product.

There has been no particular effort to distinguish between more sophisticated products and more basic ones when identifying the relevant product market, except from the heavy-handed attempt by the Court of Appeals in *US Microsoft II*. In particular, no test has been established to identify when a component of a product becomes a product in itself. *EC Microsoft* is a basic illustration of this point. Clearly, a test is needed to make such a distinction to ensure that products are not unbundled unnecessarily.

Most consumers would perceive a car to be a product, and it would probably be acceptable to them that, although part of a car, a car radio is a (separate) product, as it is also possible to purchase the car radio on its own. Equally, there are other parts of a car which can be purchased on their own, for instance when the car breaks down; these are often referred to as spare parts. Moreover, if the car will no longer drive and it is not possible to repair it, it is still possible to salvage some parts of the car. These are capable of being seen as products, despite being referred to as spare parts or parts of a car. In comparison, if a mobile phone breaks, some components or functions of the mobile phone will not be salvageable. Parts such as the battery and the SIM card would be retrievable, but if the phone no longer functions one cannot take the camera or music player out of the phone and reuse it either on its own or in another phone. In such a case, the components of the mobile phone product must be said to be truly integrated.

Taking into consideration the points above, the following test is proposed to distinguish between integrated components of a product and a (separate) product: the question is *whether the component of the product is capable of being reapplied either on its own or in another product if the original product breaks and can no longer function.*[129]

If, based on the test, the component is an integrated part of the product, one can no longer talk about tying of products, as it would be seen as one product.

---

[129]   Economists would most likely refer to this as a form of technical bundling. If for instance a PC operating system breaks down, the owner will not only lose all functions of the operating system, but also software that the owner himself has installed onto the computer. This does not make the added software an integrated part of the operating system as it is not in principle lost, but can easily be re-installed on another computer.

This broken product test therefore creates a form of safe haven, setting the boundary for when the regulator should no longer take the products apart regardless of a separate demand for the tied product. The test moves away from the consumer demand requirement and thereby the question whether there are other competitors, and instead focuses on the product itself.[130] The test could therefore be said to be more objective in comparison to the consumer demand test; on the other hand it may require some elements of technical assessment of the products, which courts may be reluctant to engage with.[131]

The broken product test will eliminate uncertainty as to when a component becomes fully integrated or when it will still be perceived as a product. However, it should be emphasised that this is all the test does. It does not say whether tying of the main product and a component that does not fall under the broken product test should be permitted. Moreover, the test is perhaps too basic to take into consideration all forms of breakdown of products, and this may mean that some components of a product may be destroyed in a break-down, but nevertheless constitute a separate product in normal circumstances. Hence, although the test offers a solution, it is not the optimal solution, as it will solve current issues but more than likely give rise to other problems.

In comparison, Areeda *et al.* assess the bundled product from a different angle, which in many ways aligns with Dolmans and Graf's perception of several levels within a market.[132] Again, the car provides an excellent example. They note that although there may be a distinct market for the component (the tied product), it does not imply that there is a separate demand for the tying product.[133] 'An independent market for carburetors does not make a car with a carburetor installed two products because no significant independent market exists for cars stripped of their carburetors . . . Two items constitute one product under the market practices test unless each could efficiently be sold without the other.'[134] The conclusion from their argument is that one cannot assess whether a component constitutes a separate product without also assessing the consumer demand for the main product without the component. If there is none then the main product *with* the component constitutes a single product, and no tie exists. One can only agree with this theory.[135]

---

[130] *US Microsoft II* at 948–949 and n. 12.

[131] Ibid., at 946, 949–950.

[132] Dolmans and Graf, *supra* n. 69, p. 228.

[133] Areeda, P. Hovenkamp, H. and Elhauge, E., *Antitrust Law, An Analysis of Antitrust Principles and their Application* (Little Brown & Company, New York, 1996), vol. X, para. 1745, p. 211.

[134] Ibid.

[135] This is also the opinion of Ahlborn, Denicolò, Geradin, and Padilla: Ahlborn, C., Denicolò, V., Geradin, D. and Padilla, A.J., 'DG Comp's Discussion Paper on Article 82: Implications of the Proposed Framework and Antitrust Rules for

Interestingly, the TTA Guidelines indicate a similar approach in respect of technology licensing, tying and bundling, and hold that products and technologies are distinct if there is a separate demand for each of the products or technology.[136] 'This is normally not the case where the technologies or products are by necessity linked in such a way that the licensed technology cannot be exploited without the tied product or both parts of the bundle cannot be exploited without the other.'[137] It is just a shame that this approach was not followed in *EC Microsoft*, because it embraces Areeda *et al.*'s theory. However, it appears to be the line of reasoning now taken by the Commission in the *Article 82 Guidance*.[138]

The CFI held instead in *EC Microsoft* that 'Microsoft's argument based on the concept that there is no demand for [Windows] without a streaming media player, amounts to contending that complementary products cannot constitute separate products.'.[139] It went on to state that although consumers only wanted operating systems with streaming media players, they perhaps wanted them from different sources.[140] The mistake the CFI made was to assume that Windows and WMP were separate products. The CFI did not assess the operating system product from an objective view, but applied Windows as the product. Therefore, although the CFI's ruling is not surprising, it is nevertheless erroneous. Moreover, it is worrying to see that a dominant company had relied upon guidance given by the Commission in its TTA Guidelines,[141] yet the reasoning within these guidelines was rejected by the CFI. This does nothing to reassure companies or create the all-important legal certainty that is essential for the effective implementation of competition law. Furthermore, it creates an unwelcoming disparity between the approach to tying under Articles 81 and 82 as well as moving the Article 82 case law approach further away from an effect-based approach.

Both *US Microsoft III* and *EC Microsoft* lead to the conclusion that the current consumer demand test must be abandoned in favour of a more forward-looking test, which can cope with the complexities of high-technology products. Areeda *et al.*'s consumer demand theory appears to offer a good starting point for such a test. The test by Areeda *et al.* is illustrated in more detail in Figure 6.1.

Dynamically Competitive Industries' 31 March 2006 (http://papers.ssrn.com/sol3/papers.cfm?abstract_id=894466), pp. 41–42.

[136] TTA Guidelines, *supra* n. 67, para. 191.
[137] Ibid.
[138] The *Article 82 Guidance*, *supra* n. 29, p. 18.
[139] Case T–201/04 *EC Microsoft*, para. 921.
[140] Ibid., para. 922.
[141] Bellis, *supra* n. 7.

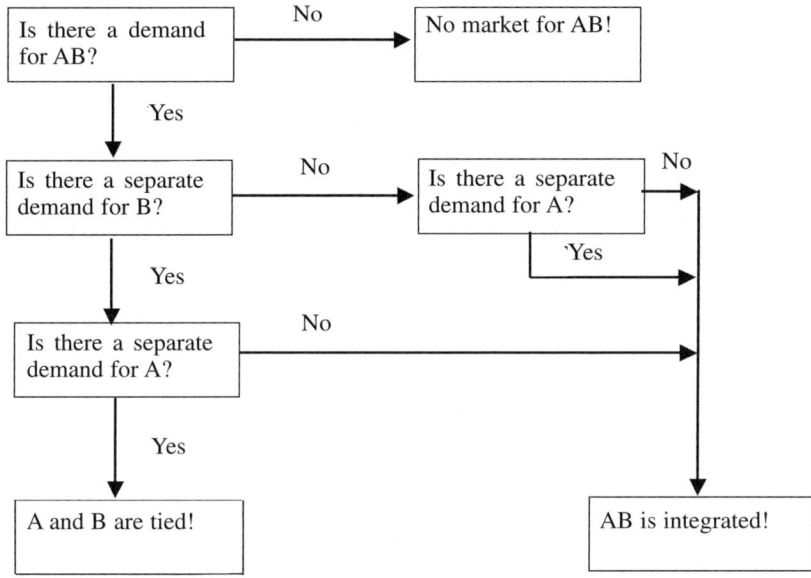

Note: Product AB consists of component A (tying product) and B (tied product)

Figure 6.1   Is there a demand for AB?

No harm to competitors will occur if there is no separate demand for either A or B, or if the products are offered in a mixed bundle. If this is not the case then there may still be an abuse as there is still is a separate demand for B, the tied product.

Areeda *et al.*'s consumer demand theory does not give a definition for demand. Does demand mean that as long as there are one or two customers demanding the product there is demand? Or should this be assessed in more realistic terms as the Supreme Court suggested in *Eastman Kodak*?[142] What Areeda *et al.*'s consumer demand theory tells us is merely that AB is acceptable as an integrated product. Although this finding is not particularly surprising, it clarifies that even if there is a separate demand for one of the components, AB can in itself be an integrated product. This is important for dominant companies, as once market power has been established; the only way a tie can be accepted is by being identified as product integration.

---

[142]   *Eastman Kodak* at 462.

Applying the figure to some well-known examples like a car (A) and its engine (B), we find that although there is a separate demand for the engine, there is not one for a car without the engine; hence a car and its engine are integrated. Attempting the same application on some case law examples such as *Hilti* there was an independent demand for the tied products, the nails (B), but unlikely to have been an independent demand for the cartridge strips (A). In comparison, in *Jefferson Parish*, where the services involved were the anaesthetic (B) and the operation (A) the services are not integrated, because there is a separate demand for the anaesthetic and there is in fact also a separate demand for an operation without the anaesthetic.

In *EC Microsoft*, the Commission and the CFI found that there was a separate demand for both Windows (A) (albeit very limited and only one particular customer group) and the WMP (B).[143] However, the CFI ignored Microsoft's argument that there was no real demand for an operating system without a media player.[144] If instead the CFI had applied the table above, together with the *Eastman Kodak* level of demand requirement, namely that the demand must be sufficient so that it is sustainable to offer the products or one of the products separately, the conclusion would have been different. This example clearly illustrates that the current consumer demand test is too rigid and cannot cope with product development and innovation. That said, just because the products have been found to be integrated does not mean they cannot cause harm.

The finding of integration does therefore not hinder a court from requiring that the integration must offer something technically different or advanced from the mere bundle of the two products, as was the case in both *US Microsoft II*[145] and *EC Microsoft*.[146] However, case law has left us in the dark as to the standard of proof on this matter. As Langer notes, the threshold of 'some advantages' suggested in *US Microsoft II* may be too lenient, but this is mainly because of the fear of strategic reasons for the technological integration not being taken into consideration.[147] If these are considered as well, the threshold may be acceptable. The CFI threshold of 'superior technical product performance' could perhaps be of better use, as the term 'superior' suggests that a significant improvement in product performance when integrated is required in comparison to the showing of some advantages. Microsoft showed

---

[143]   *EC Microsoft* Commission Decision, para. 808 and Case T–201/04 *EC Microsoft*, paras. 920–933.

[144]   Ibid., paras. 919–924.

[145]   *US Microsoft II* at 950–952.

[146]   Case T–201/04 *EC Microsoft*, para. 1159.

[147]   Langer, J., *Tying and Bundling as a Leveraging Concern under EC Competition Law* (Kluwer Law International, Alphen aan den Rijn, 2007), p. 99.

in *EC Microsoft* that Windows operated faster with WMP integrated, but the CFI dismissed this, and the 'superior technical product performance' threshold was thus not reached.[148] In comparison, such a showing may have reached the *US Microsoft II* threshold of 'some advantages'.

The standard of proof must therefore be at a level where it can be demonstrated that the technological integration clearly offers something which cannot be achieved from the mere bolting of the two products together, something 'superior', which would outweigh any potential anti-competitive effects the technological integration may generate.[149]

The Areeda *et al.* consumer demand theory does not, however, cover all situations. For instance, in the example of a mobile phone with a camera function, there would still be a demand for the phone without the camera, yet a mobile phone with a camera is seen as one product. This however, is not the case in the example of the car and the carburettor. The difference between the two components – the carburettor and the camera – is their functionality in relation to the main product. The carburettor is essential to make the car drive, whereas the camera is an additional feature of the phone, but is not (yet) essential to its functioning. Areeda *et al.*'s consumer demand theory thus requires an additional element: an assessment of the necessity of the component to the main function of the product.

Turning this argument on its head, it essentially means that when a component constitutes a necessary element for the functioning of a product, the component should never be seen as a separate product in respect of tying, regardless of whether there is a consumer demand for it on its own. This is because the demand for it is based on the presence of the main product, and the demand for the component is in essence as a spare part. This conclusion fits with the arguments given by Dolmans and Graf, who noted that there are different levels of the market and these should be considered when assessing tying and consumer demand.[150] Areeda *et al.*'s consumer demand theory identifies a simple method to establish whether products are integrated or not, although the test cannot say anything in relation to whether the integration should be permitted. This requires a further question, namely whether the integration offers a new functionality, which cannot be achieved by the consumer herself by combining the two products. If so, the products are integrated and should be permitted, regardless of the company's status on the market.

---

[148]   Case T–201/04 *EC Microsoft*, paras. 1160 and 1167.

[149]   Compare with Brief of Prof. Lawrence Lessig as *Amicus Curiae*, in *United States v Microsoft Corp.*, D.C. of Columbia, Civil Action No. 98-1233 (TPJ), 1 February 2000, pp. 11–13.

[150]   See above and Dolmans and Graf, *supra* n. 69, p. 228.

Another issue regarding the test is the level of consumer demand, as briefly mentioned above. It was seen in *EC Microsoft* that to expect the complete absence of independent demand for one of the components is an unrealistic high threshold. The US approach in *Eastman Kodak* was more sensible because it only required proof that competition was no longer efficient. However, this threshold is also high and does not sit well with innovative markets where there may still be a demand for a particular component of a product, but it is likely to cease in the future. It must therefore be asked whether there is evidence of such a process within the market when dealing with integrated products under the presumption that market participants are better placed to identify the trends and developments; if so the no consumer demand threshold has been reached. In other words, the courts must look to the market conditions.

The Areeda *et al.* consumer demand theory, together with the question of new functionalities and evidence of market development which would lead to demand ceasing for the tied product in the future, creates the expanded consumer demand test, where consumer demand is defined as sufficient demand, so it is still efficient to offer the tied product separately. The test thus encompasses the two points highlighted at the beginning of this section, namely that it covers the three perspectives, the consumers by assessing the demand, the manufacturers by looking at market developments, and finally an objective assessment of the tying arrangement by allowing the courts to assess whether the technological integration offers 'superior' benefits which are not achievable from the mere bolting of products and which outweigh any potential anti-competitive effects the technological integration may generate. Second, the test will allow the courts to distinguish between contractual tying and technological integration. This test looks to the relationship between the components, which is not dissimilar from the US misuse doctrine's distinction between staple and nonstaple products, where the latter are (often) an essential part of the product albeit not IP protected.

In comparison to the broken product test described above, the expanded consumer demand test maintains the focus upon consumer demand. Adopting this test would therefore sustain a link with previous case law and thereby create some level of legal certainty, while at the same time being more realistic and flexible towards product combinations and technological integrated products. The assessment will be based upon consumer demand – an assessment form which courts are relatively familiar with and thus perhaps more comfortable applying in comparison to the broken product test, which would require different forms of evidence and a more novel assessment method because of the need to look at the technical aspects of the product.

4.6. Recommending the Expanded Consumer Demand Test

The case law analysis undertaken in Chapters 3 and 4 identified that the consumer demand test and the relevant market assessment currently applied by the courts on both sides of the Atlantic are insufficient and underdeveloped to deal with the more complex and technologically advanced products. In particular, the current assessment is not able to distinguish between products and components of a product. Naturally, this has a significant effect on the application of the relevant market definition to tying conduct, as the latter seems to blind the courts, leading them to make hasty market definitions purely based on the presence of independent manufacturers and the scope of IP rights. In the US it was recognised in both *US Microsoft II* (indirectly) and *III* (directly) that the consumer demand test was inadequate to cope with high technology products. *US Microsoft II* offered an alternative approach and required the showing of 'some advantages' in comparison to the mere bolting of the products, whereas *US Microsoft III* rejected the use of the consumer demand test for software products, but did not give an alternative solution.

In this section, we have assessed several alternatives to the current consumer demand test and also whether we could do without it. We have found that there is a need for such a test, but, importantly, there are indeed good alternatives – in particular, the expanded consumer demand test. It was found to fit well with previous case law, and at the same time offers a relatively simple model which takes into account the relationship between the tying and tied product, the market participants' perception by assessing the market conditions, while maintaining the consumer demand as a yardstick, and is therefore easy to apply for both courts and companies. The expanded consumer demand test offers a safe haven relating to the product rather than market share. This is important for dominant companies wanting to remain innovative, because it permits a form of tying, that of technological integration. On the other hand, it was also seen that there are methods which are more sophisticated available for assessing products and markets. For instance, it is clear that the approaches taken in the TTBER and the US IP Licensing Guidelines could be adapted successfully to tying cases. However, the downside is a more complex test which may cause more harm or confusion than legal certainty. The overall recommendation is therefore the application of the expanded consumer demand test.

5.   STEP 2: DOMINANCE – ALTERNATIVE SOLUTIONS

5.1.   Introduction

There is no doubt that currently the whole tying assessment stands or falls upon whether two markets, and thereby two products, can be identified. However, what has been made clear in the above chapters is that the tying conduct in itself need not cause harm; it is other factors such as whether the company holds market power and applies this to create anti-competitive effects in the tied product market. The market power element needs to be merely the second screen of the new tying test, not the deciding factor of whether a tying arrangement is illegal.

The EAGCP report suggested removing the dominance requirement altogether from an Article 82 analysis to move to a strict effect-based approach.[151] This is firstly not in accordance with the wording of Article 82, and secondly an unnecessary intervention, because allowing for a more flexible dominance threshold, one that is more closely related to the (potential) anti-competitive effects, could equally create a more economic rule, and would at the same time provide some level of legal certainty to companies. Case law already permits for such interpretation of Article 82 when referring to super-dominance and Section 2 to monopolisation, and it is therefore a logical starting point.[152]

Whether tying is assessed from an economic or legal perspective, whether it involves IP rights or not, both economists and lawyers can agree upon the fact that tying requires market power to be harmful. However, the question is how much market power? According to the findings in Chapter 2, tying will cause harm only if the company holds *significant market power* in the tying product market such that it holds a monopoly or near-monopoly position.

US antitrust law has not answered the question of how much market power; only given vague guidance. It is thus clear that Jefferson Parish Hospital, holding a 30 per cent market share, did not possess market power or, as it has been referred to in cases, sufficient economic power,[153] yet some scholars have indicated that the market power threshold for tying is lower than that for finding monopolisation under Section 2.[154] This leaves us with an indication, in market share terms, that somewhere above 30 per cent can be sufficient to

---

[151]   It argued that where a company's conduct caused anticompetitive harm, the company was also dominant: the EAGCP Report, *supra* n. 25, p. 13.

[152]   Cases C–333/94 P, *Tetra Pak II* and C–395/96 P, *Compagnie Maritime Belge Transports SA v Commission* [2000] ECR I–1365, [2000] 4 CMLR 1076; *US Microsoft III* at 51.

[153]   *Jefferson Parish* at 13–14.

[154]   Hylton and Salinger, *supra* n. 23, p. 475.

establish market power. That said, in respect of IP rights there has been a trend under both jurisdictions to presume market power, and this has been further enhanced by the narrow market definition, as explained above. However, in the most recent US case, *Illinois Tool*, the Supreme Court for the first time clearly stated that, irrespective of a patent, the plaintiff must prove that the defendant has market power in the tying product market.[155] Hence, holding a patent should not in itself necessarily constitute market power. This conclusion is in line with section 271(d)(5) of the Patent Act,[156] which currently shapes the US misuse doctrine, as it notes that tying of patents or patented products becomes illegal only if the patentee holds market power in that market. Again, what constitutes market power is unclear.

In EC case law, we actually find more precise guidance because in case law market power has so far always been above 60 per cent. However, according to the EC Courts' general policies under Article 82, that will not render safe a company with a market share of 40 per cent which is exercising tying. There is too little case law to reach such a conclusion yet. As regards IP rights, EC case law has been rather more vague. The ECJ did state early on that an IP right would not alone constitute a finding of a dominant position.[157] Even though the EC Courts and the Commission have maintained this view, there is no doubt that IP rights have played an important role as indicator of dominance, as was illustrated in both *Hilti* and *Tetra Pak II*. It is thus clear that the alternative solutions should seek to create a safe haven for companies to generate greater legal certainty and at the same time offer a realistic assessment of the companies' market power.

5.2. The Indispensability Test

An alternative solution could be the indispensability test established in *Oscar Bronner* and discussed above, which relies on indicators other than market shares. The indispensability test in *Oscar Bronner* is fulfilled if there are technical or legal obstacles making it impossible or unreasonably difficult to produce the product.[158] Upon assessing this, one can take economic viability into consideration; however, this cannot alone constitute a valid reason, unless it is not economically viable to create a second product or service of a similar size.[159]

---

[155]   *Illinois Tool* at 16.
[156]   35 U.S.C. § 271(d)(5).
[157]   Case 78/70 *Deutsche Grammophon*, para. 16.
[158]   Case C–7/97 *Oscar Bronner*, para. 44.
[159]   Ibid., paras. 45–46.

It is clear from the outset that in this test indispensability can only be found if the company holds significant market power, and thus this test is at first glance a suitable solution. Nevertheless, there are flaws with the test. Firstly, as noted above, a successful product can often appear to be indispensable, especially if it is close to becoming a *de facto* industry standard. A successful product, however, is not necessarily indispensable: consumers may simply prefer it to alternative products.[160] Secondly, if the product is protected by IP rights there is a risk that the product will be seen as indispensable due to the IP right.[161] Whether an IP-protected product will be seen as indispensable turns on how the relevant market is defined and how IP rights are considered. In the case law, IP rights have so far received rather harsh treatment and thus presumed to convey market power.[162]

Thirdly, the indispensability test will set a very high level of market power as a precondition for illegal tying. The company would be likely to have a quasi-monopoly or monopoly in the tying market and this may make the test too lenient. For example, there may be situations where the tying product is not indispensable but is nevertheless able to create anti-competitive effects in the tied product market. Such tying arrangements will not be encompassed by the indispensability test.

EC Microsoft offers a good example of the applicability of the indispensability test. In the case, it was clear that Microsoft held market power in the operating systems market and appeared indispensable, in the sense that it had almost become an industry standard. However, there are no legal obstacles to hinder the creation of an alternative operating system. In fact, there are others in the market, albeit with small market shares. That said, due to the fact that Windows has become a *de facto* industry standard, many application programmes are created specifically for Windows. Therefore, it makes it technically difficult for a new operating system to enter the market to the extent that it is not economically viable to create a similar operating system, and certainly not one of similar size to Windows. It could therefore be argued that Windows is indispensable and tying should not take place. This is not to say that Windows as a product is indispensable, merely that Microsoft holds such a strong position on the operating systems market that tying of the operating system with another product will create significant anti-competitive effects. However, if Microsoft's market power had been lower and there were no technical obstacles hindering the introduction of a similar operating system, tying

---

160   Drexl, *supra* n. 116.
161   See Cases C–241–42/91 P *Magill* and Commission Decision *NDC Health/IMS: Interim Measures* [2002] OJ L59/18, [2002] 4 CMLR 111.
162   Cases T–30/89 *Hilti* para. 93 and T–51/89 *Tetra Pak Rausing SA v Commission* [1990] ECR II–309, para. 23.

would be permissible under the indispensable test. Yet in such a situation the test does not take into consideration all the anti-competitive effects that the tying can generate and the test would therefore allow for too many type two errors. This approach must therefore be rejected.

5.3. Market Share Cap

Another solution to the safe haven problem would be to set a market share cap. If a company falls below the cap, the company is safe and can offer tie-ins without fear of being found to infringe Article 82. Introducing a market share cap would be in line with the approach taken under Article 81 and the EC Merger Regulation.[163] However, the question it raises is what should the market share cap be? Under Article 81's block exemptions the market share cap is around 30 per cent; the EC Merger Regulation offers a safe harbour below 25 per cent. A similar safe haven for Article 82 of 30 per cent would not create any false dominant rulings. However, as was concluded in Chapter 2, it is only when the tying company holds significant market power that it will harm competition. In other words, a 30 per cent market share cap is not a sufficient safe haven: the market share cap would have to be higher – perhaps around 60 per cent – to ensure full protection of harmless tying arrangements. Setting the cap this high would be rather controversial in comparison to Article 82 case law, where in general 40 per cent seems to be the lowest denominator of dominance, and it is thus unlikely that either the Commission or the EC Courts would accept such an approach. There are nevertheless arguments for doing so. Firstly, a company holding market shares of less than 40 per cent is unlikely to be dominant in the first place, and thus unlikely to fall within the scope of Article 82 at all. Setting a market share cap lower than 40 per cent would therefore not achieve a genuine safe haven under Article 82 for companies engaging in tying. Secondly, *only* tying *in combination with* significant market power can give rise to anti-competitive effects. Tying case law suggests that significant market power is approximately 60 per cent and above.

However, market shares are only a proxy for dominance. They only provide an initial picture of a company's position in the market, and other indicators must also be taken into consideration to provide a fuller picture. A market share cap can work relatively safely with lower market share caps, but a more detailed assessment of dominance is essential to distinguish legal from illegal

---

[163] Council Regulation (EC) No 139/2004 of 20 January 2004 on the control of concentrations between undertakings [2004] OJ L24/1, recital 32; see also Commission Regulation (EC) No 2790/1999 of 22 December 1999 on the application of Article 81(3) of the Treaty to categories of vertical agreements and concerted practices [1999] OJ L336/21.

tying with higher market shares. A market share cap that high can therefore not in itself provide a realistic safe haven for pro-competitive tying arrangements. With higher market shares, a more complex test is needed to allow for safe havens.

As Monti points out, a market share cap increases the risk of under-enforcement.[164] In other words, a company may well have market power and still fall below the market share cap.[165] Nevertheless, tying arrangements can be caught under Article 81 and therefore the risk of under-enforcement is minimised, because even if the company's market shares are below the threshold its conduct can still be regulated.[166] From current case law under Article 81, it appears that the market share cap notion is working. There has been some criticism of the difficulties involved in identifying the actual market shares.[167] However, overall it seems to be a workable notion. The market share cap does not actually work alone, but in combination with other rules. A similar approach could be adopted for Article 82 tying cases.

It seems to be generally accepted that companies with market shares below 40 per cent are rarely dominant, and hence unlikely to cause harm to competition with their actions. An initial screening of a tying case could thus rule out the risk of anti-competitive effects from companies having below 40 per cent market shares.

Chapter 2 indicated that it is only when a company holds significant market power that there is a risk of harm to the market, and therefore a second screening could be introduced to take effect where the tying company holds a market share of between 40 per cent and 60 per cent. Ahlborn *et al.* pointed to two conditions in particular which, if present, would be evidence of potential anti-competitive effects.[168] The two conditions were significant market power and an imperfect competitive market. It was illustrated that the mere requirement of an imperfect competitive market was too broad a term to be meaningful in a screening of a tying case.[169] Based on this finding, it is suggested that tying can be permitted in these cases if the tied product market is sufficiently competitive, has low barriers to entry, and competitors are able to compete with the dominant company's tying product package to some degree. These were factors identified in Chapter 2 as pointers to a healthy market situation, where the tie is unlikely to cause anti-competitive harm.

---

[164] Monti (2006), *supra* n. 26, p. 47.
[165] Ibid.
[166] Ibid., p. 48.
[167] Ibid., p. 47.
[168] Ahlborn, Evans, and Padilla (2004) *supra* n. 23, pp. 331–333.
[169] Chapter 2, Section 7.

In cases where the tying company holds market shares above 60 per cent or above 40 per cent and the other factors are not fulfilled, a full assessment of the tying arrangement will be necessary to weigh anti-competitive effects against potential pro-competitive effects. Two advantages of such a market share cap test can be identified. First, the emphasis in legal analysis will shift from focusing upon the relevant market definition and establishment of dominance, as currently is the case, to an abuse analysis assessing the anti-competitive effects of the company's behaviour.[170] Secondly, the assessment of barriers to entry will fall under the abuse analysis or, in the case of tying as part of the second screening, and thus remove the problem with identifying conduct as an indicator of dominance.[171]

It should be emphasised that the market share cap test relies upon how the market is defined and thus, in respect of IP rights, if a continuation of the strict attitude towards them is maintained, there will be little if no relief in such a market share cap test. This market share cap test differs significantly from the current case-law approach by actually assessing the tying conduct from a legal *per se* perspective. This is in line with the ruling in *US Microsoft III* and the misuse doctrine,[172] the conclusions and findings of Chapters 2 and 4, but also with the recommendations made in sections 2 and 3.

5.4. Recommending the Market Share Cap

Based on the analysis above, it is recommended that the indispensability test be rejected, because it does not capture all anti-competitive effects a tying arrangement can generate, and instead a market share cap system should be introduced in the legal analysis of tying. This will create a safe haven for certain tying arrangements, screen out pro-competitive ties to establish much-needed legal certainty, and shift the balance in the tying analysis towards the anti-competitive effects assessment rather than market power, as currently is the case. Table 6.2 below illustrates how it would work.

---

[170] Monti (2006), *supra* n. 26, p. 47.
[171] Ibid.
[172] *US Microsoft III* at 92–93, but note comment at 95.

Table 6.2   Market share cap test

Market share (MS)	Action
1) MS < 40%	Tying not harmful
2) 40% < MS > 60%	Tying can be permitted in these cases if the tied product market is sufficiently competitive, has low barriers to entry and competitors are able to compete with the dominant company's tying product package to some degree.
3) 60% < MS OR 40% < MS and conditions in 2) are not fulfilled	A full assessment of the tying arrangement will be necessary to weigh anti-competitive effects against potential pro-competitive effects.

6. STEP 3: ANTI-COMPETITIVE EFFECTS – ALTERNATIVE SOLUTIONS

6.1. Introduction

Case law has shown that the anti-competitive effects test has historically been applied rather leniently with a general presumption that tying is bad. This phenomenon suggests that tying has almost been seen as illegal *per se*.[173]

In Chapter 2 it was established that there are two factors that need to be present, besides the existence of two products, before tying can cause anti-competitive effects; the company must hold market power, and there must be imperfect competition in the tied product market before the tie can be deemed harmful. This is the case if there is insufficient competition in the tied product market, and therefore high barriers to entry, and competitors are unable to compete with the dominant company's tying product package to some degree.[174] Imperfect competition has rarely been assessed by the courts on either side of the Atlantic. Instead, in the US, the anti-competitive effect assessment has focused upon whether the tying arrangement affects a substantial amount of commerce in the tied product market.[175] The requirement stems

---

[173]   See Case C–53/92 P *Hilti*, and Case 311/84 *Télémarketing*.

[174]   Chapter 2, Section 7, and Ahlborn, Evans and Padilla *supra* n. 23, pp. 331–333.

[175]   *International Salt* at 396 and *Northern Pacific Railway* at 6.

in particular from the wording of Clayton Act section 3. The Supreme Court's interpretation of this condition has been confusing. The Supreme Court initially looked upon the dollar-volume in the tied product market as an indicator of whether a substantial amount of commerce had been affected. This implied that evidence of foreclosure had already taken place was necessary, despite a low threshold adopted in *International Salt*.[176] In *Jefferson Parish*, the Supreme Court moved away from the dollar-volume assessment and instead asked whether the tie affected competition on the merits in the tied product market.[177] *Jefferson Parish* made it clear that even a risk of future foreclosure could fulfil the condition.[178] This interpretation is much more in line with the approach under EC competition law, although it does limit legal certainty.

The thorough assessment of anti-competitive effects found in *EC Microsoft* is probably more a reflection of the Commission's need to support its own conclusion that there are two separate products than a new and more robust approach to the requirement of proof of anti-competitive effects. This supposition is confirmed by the CFI's conclusions, which indicated that the comprehensive anti-competitive effects assessment undertaken by the Commission was in fact superfluous.[179] The CFI concluded that the Commission was correct in finding that the tying had a reasonable likelihood of lessening the competition in the market in the foreseeable future.[180] The conclusions by both the Commission and the CFI fail to take into account pro-competitive effects as well as the fact that high technology markets are dynamic and thus tying may be part of a natural development of the market, as was demonstrated in Chapter 2.[181]

Despite its rigid and very legalistic approach (in comparison to the more economic approach attempted by the Commission), the CFI's ruling gave some clarification of the standard of proof of anti-competitive effects necessary in tying cases. There must be a reasonable likelihood that the conduct will lessen competition within the market, so that effective competition structure is undermined within the foreseeable future, setting an important precedent of a low threshold for the standard of proof. With the exception of the Commission's thorough assessment in *EC Microsoft*, the Commission and the

---

[176] *International Salt* at 396.
[177] *Jefferson Parish* at 14.
[178] Ibid., at 16.
[179] Case T–201/04 *EC Microsoft*, paras. 857–859.
[180] Ibid., para. 1089.
[181] Mainly by Evans, D.S., Padilla, A.J., and Polo, M., 'Tying in Platform Software: Reasons for a *Rule-of-Reason* Standard in European Competition Law' (2002) 25 *World Competition*, 509, p. 509.

EC Courts have in general swept over the anti-competitive effects analysis as the low threshold could easily be reached.

It is disappointing that a requirement as important as the anti-competitive effects test has been given so little attention in tying case law. In principle, any tying arrangement should stand or fall on the basis of the anti-competitive effects it triggers. The problem, however, is that a full-blown analysis of the anti-competitive effects is difficult and creates little, if no, legal certainty. Therefore, it makes sense from a legal perspective to commence with the assessment of market power after having established the existence of two products, as market power will indicate whether the tie is likely to be harmful. This approach has also been adopted by some economists.[182] Some economists advocate a balancing test weighing up the pro-competitive effects against the anti-competitive effects, although this will require a more extensive case analysis. Others disagree and warn that such an approach has two flaws in particular. Firstly, it is time-consuming and it is difficult to weigh up real world pro- and anti-competitive conditions, some of which cannot be shown *ex ante*.[183] Secondly, the risk of regulatory errors is greater as it is mainly competitors fearing being pushed out of the market who will complain, rather than customers and consumers.[184]

That said, with the introduction of the two product assessment and the market share cap screening above, it will be only very serious anti-competitive tying arrangements that will require a full balancing analysis.[185] The following sub-sections assesses several different options of a new anti-competitive effects test, taking into consideration the flaws highlighted above in respect of the current approach and the recommendations made in the previous sections above. The first option is the indispensability test suggested by Art and McCurdy, which offers an alternative perspective on tying. The last three subsections discuss the sacrifice test, the no economic sense test, the as efficient as competitor test, and the consumer harm test. All four tests have been put forward under the Article 82 review and Section 2 debate as alternative assessments of exclusionary conduct.[186] Some thought is given to whether one

---

[182]   Kühn, Kai-Uwe, Stillman, Robert and Caffarra, Christina, 'Economic Theories of Bundling and their Policy Implications in Abuse Cases: An Assessment in Light of the Microsoft Case' (2005) 1 *European Competition Journal* 85, p. 114, and Ahlborn, Evans, and Padilla *supra* n. 23, p. 330.

[183]   Melamed, A.D., 'Exclusionary Conduct Under the Antitrust Laws: Balancing, Sacrifice, and Refusal to Deal' (2005) 20 *Berkeley Tech. Law Journal* 1247, p. 1254.

[184]   Ridyard, *supra* n. 23, p. 317.

[185]   See *US Microsoft III* at 58–59, 64, 67, 71, 74, 75, 76, 77, 78, and O'Donoghue and Padilla, *supra* n. 18, p. 192.

[186]   Gavil, Andrew I., 'Exclusionary Distribution Startegies by Dominant Firms: Striking a Better Balance' (2004) 72 *Antitrust Law Journal* 3; Salop, Steven C.,

of these tests could replace not just the current anti-competitive effect assessment step in the current tying test, but the actual test itself, and thereby be applied as a one-fits-all rule under Article 82 and Section 2. Finally some thought is also given to the standard of proof.

6.2. Indispensability Test

Art and McCurdy offer an interesting reflection as they compare the anti-competitive effect analysis of *EC Microsoft*, where the search was for actual or potential foreclosure similar to the test applied in refusal to supply cases, and in particular *Oscar Bronner*, which also dealt with access to a distribution network. [187] They do so with the postulation that the Commission was in reality not concerned about the bundling of WMP with Windows, but rather access to the OEM pre-installation distribution system over which Microsoft had control via its licensing agreements.[188] On this basis, it makes sense to see the case as a refusal to supply case rather than a tying case.[189]

As mentioned previously, *Oscar Bronner* did not involve IP rights. One could therefore question why Art and McCurdy did not choose a comparison with more IP-related cases such as *Magill* and *IMS*. However, both these cases indicate that access will be allowed only under exceptional circumstances, one of these being that the refusal to give access is hindering the launch of a new product. This is clearly not the case in *EC Microsoft*, nor was it the case in *Oscar Bronner*.

The ECJ held firstly that a refusal to provide access becomes an abuse only if:

> the refusal of the service comprised in home delivery be *likely to eliminate all competition* in the daily newspaper market on the part of the person requesting the service and that such refusal be *incapable of being objectively justified*, but also that

'Exclusionary Conduct, Effect on Consumers, and the Flawed Profit-Sacrifice Standard' (2006) 73 *Antitrust Law Journal* 311; Vickers, John, 'Abuse of Market Power', chapter 11 in Buccirossi, Paolo (ed.), *Handbook of Antitrust Economics* (MIT Press, Cambridge, Mass., 2008); Werden, *supra* n. 63.

[187]   Art, Jean-Yves and McCurdy, Gregory, 'The European Commission's Media Player Remedy in its Microsoft Decision: Compulsory Code Removal Despite the Absence of Tying or Foreclosure' (2004) 25 *European Competition Law Review* 694, p. 700.

[188]   Access to the distribution system is also an indirect access to content providers: ibid.

[189]   Compare with Langer, J., 'Bundling, A Four-Step Test to Assess the Exclusionary Effects of Bundling under Article 82 EC', Chapter 8 in Amato, G. and Ehlermann, C-D. (eds), *EC Competition Law, A Critical Assessment* (Hart Publishing, Oxford, 2007), pp. 324–325.

the *service in itself be indispensable* to carrying on that person's business, inasmuch as there is *no actual or potential substitute* in existence for that home-delivery scheme[190] [Emphasis added].

It then concluded that this was not the case as there were indeed other means of getting the newspaper to the readers, such as postal delivery and kiosk sale.[191]

As Art and McCurdy point out, there are alternatives to the OEM pre-installation. Although these may be less advantageous, according to the ECJ in *Oscar Bronner*, this is not sufficient to require access to the OEM pre-installation, which is what the Commission requires indirectly in its tying rules. Under this assessment, the indispensability requirement is not fulfilled in *EC Microsoft*. The main reason Microsoft has been successful in the OEM pre-installation of WMP is its market power in the operating system market. Hence, as the ECJ in *Oscar Bronner* held, there are in fact legal, technical and economical obstacles that make it difficult, if not almost impossible, to create a similar distribution system via its licensing agreements.[192] However, there are in fact other alternatives such as downloading and retail distribution, which are not such unfavourable options as the Commission makes them out to be. The Commission's concern was clearly with the consequences of the OEM pre-installation distribution because it feared that the market would tip in favour of the WMP.[193]

Nonetheless, there are two issues with Art and McCurdy's indispensability test. Firstly, the test can only be applied to cases with similar facts to *EC Microsoft* and *Oscar Bronner*, and not tying cases in general, as *Oscar Bronner* in effect deals with essential facilities.[194] Secondly, assessing these types of cases under the heading of refusal to supply rather than tying would raise the standard of proof from proving dominance and the potential for foreclosure effects on the tied market to instead proving indispensability of the tying product and risk of elimination of competition on the tied market. Consequently, even a powerful company like Microsoft would have the potential to escape an indispensability standard of proof.

---

[190]   Case C–7/97 *Oscar Bronner*, para 41.
[191]   Ibid., paras 42–43, see also paras 44–45.
[192]   Ibid., para. 44.
[193]   *EC Microsoft* Commission Decision, para. 878 and Hellström, Per, 'The Microsoft Judgment' Conference, 25 September 2007, BIICL, London, according to Art and McCurdy, the Commission intended to apply, a 'must carry' remedy, which would have forced Microsoft to include competitors' media players with Windows: Art and McCurdy, *supra* n. 187, p. 700.
[194]   Langer in Amato and Ehlermann, *supra* n. 187, pp. 325–26.

The ECJ's conclusion in *Oscar Bronner* was based on the concern that competitors should not be able to free-ride on rivals' competitive advantages, merely because the latter held a dominant position.[195] If it can be concluded that the Commission has applied an alternative route to ensure access to Microsoft's distribution system, then naturally the Commission has taken one step too far. Such a conclusion relies on the fact of whether the tying of the WMP and the operating system really is an abuse, and the Commission is right in its claim that the market is at risk of tipping due to this behaviour.

In any case, relying on refusal to supply essential facilities as a test for anti-competitive effects of tying assumes that the two are mutually exclusive. This is not the case. Although, in some instances tying arrangements have the same effect as refusing to supply an essential facility, the former behaviour can be much broader in its scope.[196] Therefore the indispensability test should be dismissed as an optimal solution for assessing tying arrangements.

6.3. The Sacrifice Test and the No Economic Sense Test

The sacrifice test and the closely related no economic sense test are based around the question whether the dominant company's behaviour would make economic sense but for its consequential restriction or elimination of competition or whether the company is sacrificing profit to achieve the exclusion of competition.[197] The sacrifice test has been criticised for being too focused upon the profit, leading to some anti-competitive conduct, such as reprisal abuses and non-price predation, not being caught and other pro-competitive conduct, such as investment into new products and product development, being caught.[198] Another problem is the difficulty in determining the actual sacrifice or the degree of the sacrifice leading to a subjective test and uncertainty, as a court has to look to a hypothetical market to determine the outcome of the conduct.[199] The DoJ has in later years applied the sacrifice test regularly in its Section 2 cases.[200] In

---

[195] Case C–7/97 *Oscar Bronner*, para. 47.

[196] Langer in Amato and Ehlermann, *supra* n. 189, pp. 325–326.

[197] Vickers, *supra* n. 186, pp. 423–25 and O'Donoghue and Padilla, *supra* n. 18, p. 185.

[198] Ibid., pp. 186–187;, Werden, *supra* n. 63, p. 414.

[199] O'Donoghue and Padilla, *supra* n. 18, p. 187 and Salop, *supra* n. 186, p. 358.

[200] Brief of the Appellees United States and the States Plaintiffs at 48, *United States v Microsoft Corp.* 253 F.3d 34 (D.C. Cir. 2001) *US Microsoft II*, (www.usdoj.gov/atr/cases/f7200/7230.pdf); Brief for Appellant United States of America at 30 (public redacted version), *United States v AMR Corp.* 335 F.3d 1109 (10th Cir. 2003) (www.usdoj.gov/atr/cases/f9800/9814.pdf), and Brief for the United States at 28 (public redacted version) *United States v Dentsply Int'l, Inc.* 399 F.3d 181 (3d Cir. 2005) (www.usdoj.gov/atr/cases/f202100/202141.pd). However, see the

American Airlines[201] the DoJ used the test to demonstrate that the conduct was predatory pricing and exclusionary, even though the conduct on the whole appeared profitable because it made no sense to engage in the conduct. Unfortunately, the 10th Circuit Court found that the DoJ had not reached the required standard of proof and sufficiently demonstrated the losses of the conduct,[202] thereby illustrating the criticisms made above that the test is difficult to apply.

Consequently, the US DoJ changed tactics and as *amicus curiae* suggested the use of the no economic sense test in *Trinko*[203] to refusal to supply cases.[204] In the case, the Supreme Court noted that 'the unilateral termination of a voluntary (and thus presumably profitable) course of dealing suggested a willingness to forsake short-term profits to achieve an anticompetitive end'.[205] The comment indicates a willingness by the Supreme Court to apply the no economic sense test to exclusionary conduct in more general terms.[206] In its *Section 2 Report*, the DoJ clarified that while it was rejecting the use of the profit sacrifice test to Section 2 cases, it found that in some circumstances the no economic sense test would be useful to identify exclusionary behaviour. However, it stopped short of recommending it as the sole test for Section 2, because it found the test ineffective where little cost was involved and would provide companies too much freedom to engage in harmful conduct despite low profit margins.[207]

The no economic sense test is in principle a two-pronged test, assessing firstly whether the conduct eliminates or has a tendency to eliminate competition, and, secondly, whether the conduct makes no economic sense but for that elimination; if so the conduct is exclusionary and should thus be caught under Section 2 and Article 82.[208] The test should therefore be more accurate than

---

Section 2 Report, *supra* n. 1, p. 42, where it recommends not to apply the profit sacrifice test to Section 2 cases.

[201]   *United States v AMR Corp.*, 140 F. Supp. 2d 1141 (D.Kan. 2001), *aff'd*, 335 F. 3d 1109 (10th Cir. 2003).

[202]   Pate, Hewitt R., 'The Common Law Approach and Improving Standards for Analysing Single Firm Conduct', Fordham International Antitrust Conference, 23 October 2003 (www.usdoj.gov/atr/public/speeches/202724.htm).

[203]   *Verizon Communications Inc. v Law Offices of Curtis v. Trinko LLP*, 540 U.S. 398 (2004).

[204]   O'Donoghue and Padilla, *supra* n. 18, p. 187.

[205]   *Trinko* at 409.

[206]   McDonald, J. Bruce, 'The Struggle for Standards', Section 2 Committee 'Hot Topics' Discussion, American Bar Association Section of Antitrust Law, Spring Meeting, Washington, D.C. 1 April 2004 (www.usdoj.gov/atr/public/speeches/203780.htm).

[207]   *Section 2 Report*, *supra* n. 11, p. 43.

[208]   Werden, *supra* n. 63, p. 418.

the sacrifice test in catching exclusionary abuses.[209] The test requires an assessment of both the reward (besides the elimination of competition) and cost of the conduct.[210] The conduct is considered harmful if 'it is expected to yield a negative payoff, net of the costs of undertaking the conduct, and not including any payoff from eliminating competition'.[211] Interestingly, the test does not look to the actual effect of the conduct, but the reasonably anticipated impact of the conduct in question.[212] The main reason is that actual effect may give a false picture of the conduct, whereas anticipated impact will allow for more objective economic considerations.[213] Acceptance of evidence of future effects corresponds well to current Article 82 case law such as *British Airways*.[214] However, this causes a problem in relation to the evidential burden, especially because the market and the market conditions may have changed since the company commenced the particular conduct and at the time it appeared perhaps to be an economically sensible strategy.[215] This may be evidenced by the company's business plans, yet in practice it can still be difficult to demonstrate.[216]

Vickers argues that the sacrifice test and the no economic sense test are too focused upon the intent of the conduct rather than the effects. '[While] the sacrifice test might be useful in assessing wilfulness or intent, it does not naturally yield a substantive standard of what behaviour is exclusionary.'[217]

Werden defends the no economic sense test by stressing that the focus of the test is not on the intent; despite the need to assess the anticipated impact of the conduct, the test will also catch exclusionary conduct, even if the company did not intend to exclude competitors.[218]

In respect of tying, some have argued that tying could and should be viewed as a form of predation;[219] the no economic sense test would therefore in principle be ideal to apply to tying arrangements. However, one should be cautious of the role that intent can play in the no economic sense test, which may be greater than is sensible for the assessment of tying arrangements. Werden's

---

[209]   O'Donoghue and Padilla, *supra* n. 18, p. 188.
[210]   Werden, *supra* n. 63, p. 416.
[211]   Ibid.
[212]   Ibid.
[213]   Ibid.
[214]   Case T–219/99 *British Airways v Commission* [2003] ECR II–5917, [2004] 4 CMLR 1008, para. 293.
[215]   O'Donoghue and Padilla, *supra* n. 18, p. 188.
[216]   Ibid.
[217]   Vickers, *supra* n. 186, p. 424.
[218]   Werden, *supra* n. 65, pp. 416–417 and 426.
[219]   Tirole, Jean, 'The Analysis of Tying Cases: A Primer' (2005) *Competition Policy International* 1, p. 26.

solution to this problem is to exclude the introduction of new products and price cuts, which remain above costs from the no economic sense test, because he is aware that applying the no economic sense test to these types of conduct risks stifling innovation and otherwise pro-competitive behaviour.[220] The argument is interesting, because it carves out a safe haven for certain tying arrangements, in particular technological integration, with a presumption that these will contain pro-competitive effects which will always outweigh the anti-competitive effects. However, this raises two concerns. First, there is the question when the products in question are integrated or two separate products merely bundled together. Thus, if a safe haven needs to be created for these types of products an initial screening must be introduced to identify these situations. This supports previous findings and recommendations above, but also creates a hole in the argument favouring a one-fits-all rule. The second concern in respect of the creation of a safe haven is that a powerful company such as Microsoft could go free when engaging in technological integration, unless some screening is introduced. In fact, some argue that even with the application of the no economic sense test (or the sacrifice test for that matter) Microsoft's conduct in integrating Internet Explorer with its operating system, as was the case in the US, would not be caught under that test, because of the small cost involved to Microsoft in integrating the two products.[221] Microsoft could make the integration at little cost precisely because of its strong market power in the operating system, yet the no economic sense test would be unable to find such behaviour exclusionary.[222] Another concern with applying the no economic sense test and the sacrifice test to tying is the requirement of recoupment or potential recoupment, but, as Langer notes, there is no need for recoupment for a tying arrangement to be profitable, and therefore harmful arrangement would not necessarily be caught by the tests.[223] Based on this finding together with the fact that the tests do not account for costs or likely errors, and do not consider the nature of the behaviour or previous experiences of such practices, Langer argues in favour of rejecting these tests as alternatives to the current approach to tying under Article 82.[224]

Although from a more general perspective O'Donoghue and Padilla also come to the conclusion that neither the sacrifice test nor the no economic sense

---

[220] Werden, *supra* n. 63, pp. 414–415.
[221] Jacobson, Jonathan M. and Sher, Scott A., '"No economic sense" Makes No Sense for Exclusive Dealing' (2006) 73 *Antitrust Law Journal* 779, pp. 794–795 and Gavail, *supra* n. 183, p. 57.
[222] Ibid.
[223] Langer in Amato and Ehlermann, *supra* n. 189, p. 324;, see also O'Donoghue and Padilla, *supra* n. 18, p. 185.
[224] Langer in Amato and Ehlermann, *supra* n. 189, p. 324.

test brings any revolutionary assessment to the table, which can clearly distinguish exclusionary behaviour from that of competition on the merits.[225]

Taking these points into consideration, neither the sacrifice test nor the no economic sense test appears to be the right solution for tying arrangements. That said, the main criticism of the no economic sense test in particular appears to be its take it or leave it nature. If, on the other hand, the no economic sense test were combined with safe havens, such as the one suggested by Werden of excluding new products, and other requirements, such as the need to establish two products to minimise both over- and under-enforcement, then the test offers a much more economic approach to the assessment of the pro- and anti-competitive effects of tying than the one currently applied. Its downsides are the recoupment requirement, the difficulty of applying it, and providing the evidence. Another problem is the fact that it is focused upon the effects upon the dominant company rather than the effects upon the consumer and the competitive process, which undeniably is important in the assessment of tying because tying is not always about production costs, but can be about product quality and choice. The sacrifice test and the no economic sense test are therefore rejected as optimal anti-competitive effects tests for tying arrangements.

6.4. As-Efficient-As Competitor Test

The as-efficient-as competitor test relies on the assumption that exclusionary conduct is only that which will exclude a competitor from the market which is equally or more efficient in relation to the dominant company exercising the particular conduct.[226] Conduct which will exclude less efficient competitors is permitted, based on the view that these competitors would not survive the competitive process anyway.[227] Cases such as *AKZO*[228] and *Oscar Bronner* speak in favour of the application of this test to Article 82 abuses, because in both the ECJ took the equally efficient competitor into account and, more importantly, Advocate General Jacobs clarified in *Oscar Bronner* that the main aim of Article 82 was not the protection of the competitors, but the competitive process and the consumers.[229]

---

[225]   O'Donoghue and Padilla, *supra* n. 18, pp. 188–189 – see also Vickers, *supra* n. 186, p. 426.

[226]   Monti (2007), *supra* n. 44, p. 214.

[227]   O'Donoghue and Padilla, *supra* n. 18, p. 185.

[228]   Case 62/86 *AKZO-Chemie v Commission* [1991] ECR I–3359, [1993] 5 CMLR 215.

[229]   Opinion of AG Jacobs in Case C–7/97 *Oscar Bronner GmbH & Co KG v Mediaprint Zeitungs- und Zeitschriftenverlag GmbH & Co KG* [1998] ECR I–7791, [1999] 4 CMLR 112, para. 58.

In comparison to the no economic sense test, there is no need to consider recoupment, and the test, according to Monti, permits the establishment of abuse more easily, but more importantly it is more in line with current EC competition law principles as it is based on foreclosure,[230] an element is of great importance to the assessment of tying on both sides of the Atlantic. Yet, in practice it may be difficult to measure comparative efficiency of the different companies in the market.[231] Moreover, less efficient companies can in fact enhance consumer welfare with their presence in the market when the increased competition outweighs the relative inefficiencies of the companies.[232] This could be the case in markets with emerging competition by new rivals or products, and there is a risk that in these markets the use of the as-efficient-as competitor test could lead to under-deterrence.[233]

In relation to tying the assessment of efficiency needs to look beyond the boundaries of the individual market, because a competitor may be able to compete efficiently with the dominant company in one market, but the dominant company may appear more efficient because of the competitive advantage it may have gained from tying two products together. When dominant companies tie, potential competitors are required to enter two markets at the same time to compete effectively, which increases the barriers to entry. In such cases it is difficult clearly to define equally efficient.[234] Yet, it is an important indicator of whether the tying arrangements is harmful, as was demonstrated in Chapter 2 and highlighted again in section 5 above with the introduction of the market share cap. The as-efficient-as competitor test may therefore be helpful in identifying when a market is sufficiently competitive with low barriers to entry and allows competitors the possibility to compete to some extent with the dominant company's tied product package.

Although, the as-efficient-as competitor test provides a relative simple economic method of identifying exclusionary conduct, it is not refined enough to cope with the more advanced forms of abuses and non-market abuses, and would in some cases lead to a complex balancing act between the harm the conduct causes to equally efficient companies and the efficiencies that the conduct does create.[235] The *Section 2 Report* concludes that the as-efficient-as competitor test may be suitable to certain pricing practices, but dismisses its

---

230   Monti, *supra* n. 43, p. 214.
231   Gavil, *supra* n. 186, p. 59.
232   Vickers, *supra* n. 186, p. 427; Salop, *supra* n. 186, pp. 328–329; Gavil, *supra* n. 186, p. 59; and O'Donoghue and Padilla, *supra* n. 18, p. 189.
233   *Section 2 Report*, *supra* n. 1, p. 44.
234   O'Donoghue and Padilla, *supra* n. 18, p. 191 – see also Gavil, *supra* n. 186, p. 59.
235   O'Donoghue and Padilla, *supra* n. 18, pp. 190–191.

use for other forms of conduct, including tying.[236] Economists such as Monti, O'Donoghue and Padilla are of the opinion that nevertheless this test is the most promising economic test for Article 82.[237] However, as Gavil notes, the as-efficient-as competitor test's centre of attention is wrongly focused; it should not purely be about the dominant company and its efficiency, but also about the efficiencies or inefficiencies of the alleged abusive conduct.[238] This is crucial to tying because the actual tying arrangement will offer the products in different packages or as an integrated product. The as-efficient-as competitor test will in effect only measure the reduced costs for the dominant company in offering the products together (if there is such a reduction) and the barriers to entry for the potential competitors, but not the benefits gained by the consumers from the tying arrangement, whether this is in the form of a new product or functionality, or a reduction in their purchase or search costs.

The as-efficient-as competitor test is therefore not as fine-tuned as would be preferred to deal with the often complex nature of the tie and the involvement of two markets. However, the test is crucial to assessing whether the tying arrangement is harmful, because it identifies whether (potential) competitors are excluded from the market due to the tie or due to efficiencies. The test therefore offers a good alternative to the current approach, and would tie in well with the market share cap system suggested above.

6.5. The Consumer Harm Test – Re-balancing the Standard of Proof

A third test, which is often discussed in relation to the no economic sense test and the as-efficient-as competitor test, is the consumer harm test. As the name indicates, it looks to the effect the conduct has upon the consumer. The question is whether consumers are harmed by increased prices, lower output and quality, or diminishing innovation.[239] In comparison to the sacrifice and no economic sense tests, the focus of this test is not on the impact the conduct has upon the dominant company and the economic motivation for the conduct, but rather on the effect the conduct has upon the market, the consumer and the competitive process.[240] The test seeks to weigh up the pro- and anti-competitive effects of the conduct and will restrict the conduct if there is a net harm to consumer welfare taking into account all relevant available information.[241]

---

[236] *Section 2 Report*, *supra* n. 1, p. 45.
[237] Monti (2007), *supra* n. 43, p. 214 and O'Donoghue and Padilla, *supra* n. 18, p. 191.
[238] Gavil, *supra* n. 186, p. 59.
[239] Salop, *supra* n. 186, p. 331.
[240] Ibid.
[241] O'Donoghue and Padilla, *supra* n. 18, p. 191.

There appear to be two approaches to the assessment of efficiencies under the consumer harm test: one is a straightforward weighing of the inefficiencies against the efficiencies,[242] and another which has been referred to as a proportionality test containing two elements: first, it must be assessed whether the conduct is reasonably capable of creating or maintaining monopoly power by restricting competitors and, second, whether the conduct is unnecessary to the consumer benefits sought to be achieved or whether it is disproportionate to the benefits the conduct brings.[243] Importantly, the DoJ rejects the use of the consumer harm test when the method of precise balancing of effects is applied (referred to in the *Section 2 Report* as the effects-balancing test) because of its 'open-ended nature' and the 'inherent uncertainty for business in predicting its outcome'.[244] Instead preference is given to the measuring of consumer harm under a (dis)proportionality test, despite the higher burden of proof this test places upon the plaintiff.[245] Hence, conduct will be seen as anti-competitive under Section 2 if 'its likely anti-competitive harms substantially outweigh its likely precompetitive benefits'.[246]

Both US and EC case law have made use of the consumer harm test, in particular in the *Microsoft* cases. In *US Microsoft III*, the Court of Appeals laid down the following steps as the approach to exclusionary conduct under Section 2. First, the plaintiff must demonstrate that the conduct harms the competitive process and thereby consumers.[247] Interestingly, the Court of Appeals indicated that the standard of proof required from the plaintiff may vary, depending upon whether the plaintiff is a private party or the government.[248] Second, if the plaintiff is successful, the defendant is hereafter permitted to submit a pro-competitive justification for the conduct.[249] Third, the plaintiff then has the opportunity to rebut the pro-competitive justification or demonstrate that the efficiencies gained from the conduct are outweighed by the harms caused by the conduct.[250] The Court of Appeals noted that this latter step was very similar to the *rule of reason* applied under Section 1. On the facts, there was little need to weigh the pro-competitive effects against the harms as Microsoft's behaviour was either clearly pro-competitive or clearly

---

[242]   Ibid., p. 192 and Gavil, *supra* n. 186, p. 61.
[243]   Areeda and Hovenkamp, *supra* n. 133, para. 651a at 72; Salop, *supra* n. 186, p. 333; O'Donoghue and Padilla, *supra* n. 18, p. 192.
[244]   *Section 2 Report, supra* n. 1, p. 38.
[245]   Ibid., p. 46.
[246]   Ibid., p. 45.
[247]   *US Microsoft III* at 58–59.
[248]   Ibid.
[249]   Ibid. at 59.
[250]   Ibid.

harmful.[251] In comparison, the Commission attempted a more thorough balancing of the pro- and anti-competitive effects in the tying section of the *EC Microsoft* case.[252] Moreover, the weighing of pro- and anti-competitive effects is commonly applied in other areas of EC competition law such as merger regulation and cooperative agreements; in particular the exemption under Article 81(3) requires a certain balancing of efficiencies and harms that the agreement will generate.[253]

The main criticism of the consumer harm test is the actual balancing test.[254] Although it works under the merger regulation and for agreements under Article 81, there is one major difference: these decisions to merge or enter into an agreement are decisions about the future, and therefore can more easily be adapted to the surrounding market conditions to ensure that the pro-competitive effects of the merger or agreement will outweigh the anti-competitive effects. This is not an option available to the same extent for unilateral conduct, and individual companies will face greater difficulties controlling this.[255] More importantly, in respect of unilateral conduct the assessment often arises during or after the alleged abusive behaviour has commenced, and therefore the burden will be placed upon the plaintiff to demonstrate that the conduct will in the (near) future cause harm.[256] In other words, although theoretically the consumer harm test is dealing with the right issues, in real world terms it may be the most difficult one to apply also from an administrative perspective in comparison to the no economic sense and as-efficient-as competitor tests discussed above.[257]

Although many economists and the *Section 2 Report* argue against the application of the consumer harm test, there are a few points which weigh heavily in its favour. First, it is already a recognised method in case law, and its application would therefore relatively easily ensure alignment with previous case law and thereby provide some level of legal certainty. Second, it would from an EC competition law perspective bring the Article 82 assessment into line with both the Merger Regulation and the modernised Article 81

---

[251] Ibid. at 64, 67, 71, 74, 75, 76, 77, 78, and O'Donoghue and Padilla, *supra* n. 18, p. 192.

[252] *EC Microsoft* Commission Decision paras. 956–970.

[253] Commission Notice: Guidelines on the assessment of horizontal mergers under the Council Regulation on the control of concentrations between undertakings [2004] OJ C31/5, paras 80–88 and Commission Notice: Guidelines on the Application of Article 81(3) of the EC Treaty [2004] OJ C101/97, para. 24.

[254] *Section 2 Report, supra* n. 1, p. 38.

[255] O'Donoghue and Padilla, *supra* n. 18, p. 193.

[256] Ibid.

[257] Ibid., p. 194; *Section 2 Report, supra* n. 1, p. 37; and Melamed, *supra* n. 183, p. 1254.

approach. From an US perspective it would correspond to the approach in *US Microsoft III*, and applying a proportionality test in relation to the standard of proof would in fact align it with the recommendations made in the *Section 2 Report*.[258] Third, the consumer harm test would not be working on its own, but would be used in conjunction with a set of screens,[259] notably for tying, the two products requirement (in the new approach the expanded consumer demand test), establishment of significant market power (the market share cap), and the as-efficient-as competitor test, before the consumer harm test comes into play. This should eliminate the concerns some economists have about the test not containing clear and concise rules.[260] It also means that the consumer harm test is not applied in the traditional American *rule of reason* manner, but much more in a traditional EC competition law approach. That means that once the consumer harm test comes into play there is either a clear finding, if the company has between a 40 per cent and a 60 per cent market share, that competitors have been excluded or, if the company has 60 per cent or above, a presumption that the market power is so significant that the conduct is exclusionary. It is with these prior findings in mind that the consumer harm test is set to work. Fourth, the consumer harm test focuses on the effects upon the consumer and the market, in comparison to the no economic sense test and the sacrifice test, this viewpoint aligns the assessment of anti-competitive effects with that of the two product requirement, where the expanded consumer demand test was recommended as the two separate prod-ucts assessment, which focuses upon the consumers and their perception of the product(s) but also the market conditions. It is therefore sensible to adopt an anti-competitive effects assessment test, which is equally focused upon consumer welfare. Fifth, even in cases where there is a need to weigh the effi-ciencies against harms, it may not always be a very difficult exercise, as US *Microsoft III* demonstrated,[261] because the benefits or harms may be so signif-icant that it is clear to which side the scale is tipping. It may therefore only be in a few cases where a more thorough balancing analysis will have to take place.

It is clear that the consumer harm test offers an alternative approach to tying arrangements, which not only allows abusive tying to be caught, but will also exempt the tying of products where there are significant benefits to consumers gained from the tie. The consumer harm test's downside is its balancing test, which makes its application difficult. The consumer harm test, however, will only be applied when either the company has been found to hold

---

258   *Section 2 Report, supra* n. 1, p. 90.
259   See Gavil, *supra* n. 186, p. 64.
260   O'Donoghue and Padilla, *supra* n. 18, p. 194.
261   *US Microsoft III* at 58–59, 64, 67, 71, 74, 75, 76, 77, 78.

market power and there is evidence of some exclusion of equally efficient competitors, or it has been found to hold significant market power, where a presumption of exclusion of equally efficient competitors is made. The consumer harm test allows for additional evidence of anti-competitive effects to be shown, and only where this can be outweighed by pro-competitive effects will the tie be permitted. The consumer harm test is therefore recommended. It demonstrates the importance of taking the benefits and efficiencies of the tie into consideration, and some form of balancing act must be made between the anti-competitive effects and the benefits generated from the tying arrangement. It therefore means that one cannot talk about an anti-competitive effects test without also discussing objective justifications, and the two tests should therefore be combined and seen as two halves of one test.

6.6. Standard of Proof

Monti notes in relation to the consumer harm test the problem of the standard of proof.[262] However, the level of proof will be relevant for whichever anti-competitive effects test chosen and therefore deserves to be treated separately. The choice of standard spans from demonstrating actual harm, all the way to demonstrating probable harm. A high standard of proof, such as actual harm would lead to too many false negatives, whereas a too low standard would lead to too many false positives.[263] Case law has demonstrated that although the standard of proof of anti-competitive effects is low, there is already in place a form of balancing test between pro- and anti-competitive effects of the assessed conduct.[264] Currently, this balancing test is not very effective or fair from a dominant company's point of view, nor realistic from an economic point of view, because any anti-competitive effects will always outweigh any pro-competitive effects raised under the objective justifications. This is due to the low standard of proof for the former, and a high standard of proof for the pro-competitive effects.[265] Two approaches to this re-balancing have been suggested: one focusing on proportionality and another focusing on preponderance,[266] the latter allowing for the conduct when the pro-competitive effects outweigh the anti-competitive effects, whereas the proportionality test will permit the conduct only if the anti-competitive effects are not disproportionate

---

[262] Monti (2007), *supra* n. 43, p. 214.
[263] Ibid.
[264] Case C–95/04 *British Airways v Commission* [2007] ECR I–2331, [2007] 4 CMLR 22, para. 86; *International Business Machines Corp. (IBM Corp.) v United States*, 298 U.S. 131 (1936) and *US Microsoft III*.
[265] Chapters 3 and 4.
[266] Gavil, *supra* n. 186, p. 64.

to the pro-competitive effects.[267] The proportionality test suggests a higher standard of proof than the preponderance test.[268] The choice of the standard of proof will at the end of the day be a matter of policy choice. In respect of EC competition law, guidance upon the direction of such choice can be obtained from the balancing test already available under Article 81(3). Under US antitrust law, the Court of Appeals held in *US Microsoft III* that the conduct should not unreasonably restrain competition, and the *Section 2 Report* recommends that only where there are no pro-competitive benefits from the tie or if anti-competitive effects are substantially disproportionate to the benefits should the tying arrangement be deemed illegal.[269]

It is suggested that the standard of proof is re-balanced, by increasing the standard of proof for anti-competitive effects while at the same time lowering the standard of proof for pro-competitive effects to the point where they are equally weighted.[270] The challenge is of course that the elimination of competitors is extremely harmful to competition and ultimately consumers, and therefore must be matched with an equally powerful pro-competitive benefit. This may be achieved from sheer efficiencies, but this is not a possibility in tying cases. It could be speculated that where the tying results in the development of products that creates a new product or generation of products, the benefit of tying may outweigh a potential to eliminate competition. This form of re-balancing is closely related to that of the proportionality test.

Importantly, the disparity in the balancing test would not actually cause such severe concerns if it was not for the narrow approach adopted to the relevant market definition, which catches (almost) all tying arrangements. Under the current model once market power and two products has been established the tying arrangement will undoubtedly be found to infringe regardless of its actual effects. If however, the new approach to tying was introduced tying arrangements would, firstly, be *legal per se*, which means only the harmful tying arrangement would be caught and would require assessment under the anti-competitive effects test.

---

[267]   Ibid. See the *Section 2 Report*, which favour the proportionality test and dismisses the consumer harm balancing test: *supra* n. 1, pp. 36–38 and 45–46.

[268]   Gavil, *supra* n. 186, p. 64.

[269]   *US Microsoft III* at 95 and the *Section 2 Report*, *supra* n. 1, p. 90.

[270]   Kallaugher and Sher suggest the same in respect of rebates; Kallaugher, J. and Sher, B. 'Rebates revisited: Anti-competitive Effects and Exclusionary Abuse under Article 82' (2004) 25 *European Competition Law Review* 263, p. 284.

6.7. Recommending the As-Efficient-As Competitor Test and the Consumer Harm Test

In finding an alternative approach to the assessment of anti-competitive effects consideration should be paid to the two points mentioned at the beginning of this chapter, namely, the question whether a rule for all abuses should be found or separate rules for different abuses should be sought. The above analysis demonstrates the difficulties in finding a test which can accommodate all types of conduct equally well.[271] The second point highlighted the importance of breadth as well as sophistication within the test to ensure that no harmful conduct escapes, but also permits efficiencies and benefits to consumers and the competitive process to be given appropriate consideration. The indispensability test relied too heavily on specific circumstances of particular cases to be a useful test for tying in general. The sacrifice test and the no economic sense test had some elements of attraction, but were focused upon the economic motivation of the dominant company and the impact the conduct would have upon it rather than the effects upon the consumers and the market. However, the examination of these tests made it clear that it will be extremely difficult to identify one rule which can apply to all exclusionary behaviour without the introduction of some screens to adjust the rule to the specific behaviour.

The as-efficient-as competitor test was found to work well in conjunction with the market share cap recommended above. It would allow for the possibility of identifying those situations where competitors may be equally efficient, but are nevertheless unable to compete sufficiently with the dominant company due to the competitive advantage it has created for itself with the tying arrangement. The as-efficient-as competitor test is recommended as a first anti-competitive effects screen and should be applied when the market share of a company is above 40 per cent, as suggested in the market share cap system. If exclusion of equally efficient competitors is found a full blown assessment of the pro- and anti-competitive effects should be made in line with the consumer harm test. If the company holds more than a 60 per cent market share, there is a presumption of exclusion of equally efficient competitors when the assessment of pro- and anti-competitive effects are made, and this means that the noose around the tying arrangement has at this point already been tightened so to speak, and therefore that any objective justification for the tie must be very strong.

---

[271] See also Vickers, *supra* n. 186, p. 430; the *Section 2 Report, supra* n. 1, pp. 46–47 and the *Article 82 Guidance* provides no indication of such initiative.

With this presumption of exclusion of equally efficient competitors it makes sense to have a more flexible anti-competitive effects test. The application of the consumer harm test in this context is therefore more in line with the traditional Article 82 approach to anti-competitive effects and the moderate *per se* rule to tying arrangements under Section 1 rather than a full blown *rule of reason* approach applied under Section 1 (and 2) to certain forms of conduct.

The as-efficient-as competitor test and the consumer harm test complement each other well, as the former looks to the effect the tying arrangement has upon the defendant and its competitors, whereas the latter focuses upon the effects of the consumers and the competitive process – a strong point in the consumer harm test's favour. On the other hand, it is without clear and concise rules and thus unlikely to create legal certainty based on sound economic principles, one of the main factors for a reform of Article 82 and Section 2, and thus also a great influence in identifying a new approach to tying.[272] However, the consumer harm test teaches us an important lesson: anti-competitive effects cannot be discussed without also considering the pro-competitive effects. The anti-competitive effects test must therefore be seen as one half of a greater test together with the assessment of objective justifications.

This is further emphasised by the discussion of standard of proof, which clarified that if the two are not aligned one will always outweigh the other, as is the situation at the moment in tying cases where the low anti-competitive effects threshold ensures that it is impossible to justify a tying arrangement regardless of any benefits generated. The recommendation is therefore that for companies holding a market share of between 40 per cent and 60 per cent the as-efficient-as competitor test is applied to assess whether the tie hinders equally efficient competitors from competing with the dominant company and thereby creates imperfect competition on the tied product market. If this is indeed the case, any anti-competitive effects should be weighed against any objective justification given for the tie under the consumer harm test. For companies holding 60 per cent or above it is assumed that equally efficient competitors are excluded to some extent from the market, and the assessment moves to the consumer harm balancing test. Only tying arrangements with a strong defence would be permitted under the new approach to tying. In other words, the defence must be in proportion to the harm caused and vice versa.

---

[272] See Vickers, *supra* n. 186, p. 430 and Commission, Article 82 Review (http://ec.europa.eu/comm/competition/antitrust/art82/index.html.), and the *Section 2 Report, supra* n. 1, p. vii.

7.   STEP 4: OBJECTIVE JUSTIFICATIONS – ALTERNATIVE SOLUTIONS

7.1.   Introduction

This section looks at different approaches to the objective justifications and defences of tying in order to identify alternative solutions. The least developed element in a tying analysis is the actual defences available for the company applying the conduct, that is despite the fact that economists have argued that tying can have both pro- and anti-competitive effects and that tying is a common business practice.[273] Neither US legislation nor EC legislation allows for exemption as such, although Article 82(d) specifically states that tying can be an abuse if it ties products that 'by their nature or according to commercial usage, have no connection with . . .' the tying product. Article 82 therefore allows for an exemption of tying where this is due to the nature of the products or the commercial usage of the products. However, case law has shown this part of Article 82 to be ineffective.

Both jurisdictions have allowed defences to be argued before the courts, such as health and safety, product quality and reduction in costs.[274] Under both US antitrust law and EC competition law the threshold has been set very high, and most defences put forward by defendants have been rejected.[275] Case law indicates that for the tying to be accepted it should be the least restrictive measure in order to achieve the justification given for the tie.[276] Only where special market conditions have been present has tying been permitted.[277]

It was made clear in *EC Microsoft* that the defences could be added together as a cluster of defences, which the Commission then required to outweigh the anti-competitive effects identified in the anti-competitive effects assessment.[278] In principle, such a balancing test is exactly what economists have advocated; however as the test currently stands under EC competition law, it is in practice worthless as the standard of proof in the two tiers of the test is different, to the extent that the anti-competitive effects will always outweigh the pro-competitive. *EC Microsoft* provides a clear example of that.

---

[273]   See Alhborn, Evans and Padilla (2004) *supra*, n. 23; Evans, Padilla, and Polo, *supra* n. 181; and Evans and Salinger, *supra* n. 23.

[274]   Note that an IP right cannot be applied as a defence for the tying arrangement.

[275]   *IBM Corp.* at 139–140; *International Salt* at 396–398; *Hilti* Commission Decision, paras 87–93; Case T–83/91 *Tetra Pak II*, paras 138–140.

[276]   *IBM Corp.* at 139–140 and *Hilti* Commission Decision, para. 90.

[277]   *United States v. Jerrold Electronics Corp.*, 365 U.S. 567 (1961) at 555–556 or a more thorough assessment was required by the Court: *US Microsoft III* at 95.

[278]   *EC Microsoft* Commission Decision, para. 958.

It was found in Chapter 2 that tying is a common business practice, and thus the defence already built into Article 82(d) would be a good initial starting point for the justification of a tying arrangement. It was illustrated in the empirical evidence section of Chapter 2 that tying is also very much a child of the industry in which the company operates, and thus consideration must be given to this fact. In this section two approaches to objective justifications will be discussed: the replication of the Article 81(3) exemption and the reintroduction of the Article 82(d) exemption of 'according to nature or commercial usage'.

7.2. Replicate Article 81(3) Exemption

In the *Article 82 Guidance* the Commission applies a more lenient approach to objective justifications than has been seen in case law. In fact it is a mirror image of the Article 81(3) exemption. Exclusionary conduct can thus be exempted under Article 82 if the dominant company can demonstrate that 'its conduct is objectively necessary' or 'its conduct produces substantial efficiencies which outweigh any anticompetitive effects on consumers'.[279] This means that even if a form of conduct is found to have some anti-competitive effects, it may nevertheless be allowed. The Commission further explains that the burden of proof lies with the dominant undertaking to provide the necessary evidence to demonstrate that the conduct is objectively justified. However, the burden of proof is then upon the Commission to demonstrate that the evidence and arguments cannot prevail and the justification cannot be accepted.[280] This corresponds with both the view taken in US cases and Article 2 of Regulation 1/2003,[281] but is a change in attitude from the Commission's approach in *EC Microsoft* and its *Discussion Paper*, where it attempted to place the entire burden upon the dominant company.[282]

The efficiency defence is based on the principle of weighing the positive results of the alleged abusive conduct against the negative ones, also called a proportionality analysis, and mirrors, as noted, the Article 81(3) exemption. It is no surprise to find it also as an efficiency defence under Article 82 in the

---

[279]   The *Article 82 Guidance*, *supra* n. 29, p. 12.

[280]   Ibid., and see Case T–201/04 *EC Microsoft*, para. 1144.

[281]   *US Microsoft III* at 95 and Council Regulation 1/2003 on the implementation of the rules on competition laid down in Articles 81 and 82 of the Treaty [2003] OJ L1/1-25.

[282]   *EC Microsoft* Commission Decision, paras 955–970 and DG Competition, Commission, 'DG Competition discussion paper on the application of Article 82 of the Treaty to exclusionary abuses', Brussels, December 2005 (http://ec.europa.eu/comm/competition/antitrust/art82/discpaper2005.pdf), p. 24.

Article 82 Guidance.[283] The Commission's suggestion is welcomed, since having consistent interpretation and similar defences across the different competition rules would create legal certainty and be helpful for the companies undertaking self-assessment before acting.[284] The Commission lists four conditions which must be fulfilled:

(1)   that efficiencies are realised or likely to be realised as a result of the conduct;
(2)   the conduct is indispensable to realisation of the efficiencies;
(3)   that the likely efficiencies outweigh any likely negative effects on competition and consumers welfare;
(4)   the conduct does not eliminate effective competition, by removing actual or potential competition.[285]

As an example of efficiencies that can be justified the Commission lists technical improvements in the quality of goods, or a reduction in production or distribution costs.[286] The second condition requires that the dominant undertaking demonstrate that the conduct is indispensable, meaning that it is the only possible option to achieve the efficiencies sought. According to the *Discussion Paper* this includes showing that the method is the only economically viable one, which is also the least anti-competitive one.[287]

The third condition should in principle ensure unnecessary restraints on competition, but also allow consumers to gain a fair share of the efficiencies. This is one of the key points in these proportionality analyses, notably because 'the Community competition rules protect competition on the market as a means of enhancing consumer welfare and of ensuring an efficient allocation of resources'.[288] Naturally, if the benefits of the conduct cannot be passed on to consumers the efficiency defence falls apart and the conduct will not be allowed. Benefits to consumers would be improved quality of products, new

---

[283]   A similar wording has been applied in Article 2(1)(b) of the EC Merger Regulation, *supra* n. 163, outlining facts that should be considered in a merger appraisal: 'the development of technical and economic progress provided that it is to consumers' advantage'. It can be applied as an efficiency defence for mergers, as highlighted in the Commission Notice: Guidelines on the assessment of horizontal mergers under the Council Regulation on the control of concentrations between undertakings [2004] OJ C31/5, paras 76–88 (the HMG). The efficiency must be passed on to consumers: para. 79.

[284]   See Guidelines in the application of Article 81(3), *supra* n. 253.

[285]   The *Article 82 Guidance*, *supra* n. 29, pp. 12–13.

[286]   Ibid.

[287]   Commission *Discussion Paper*, *supra* n. 282, pp. 26–27.

[288]   Ibid., p. 27.

innovative products and lower prices; however, if the efficiencies are in the form of cost savings, the consumers may not receive a fair share because, as Rousseva points out, this condition relies on competition still remaining in the market, and that of course has been weakened by the presence of the dominant company.[289] Whether a fair share is passed on to the consumer should be assessed in relation to the time it will take for these benefits to reach consumers.[290] The later the efficiencies reach the consumer the less significance should be placed upon them.

Finally, the fourth condition – ensuring that competition is not eliminated as a result of the conduct – is closely related to the consumer benefit requirement. The dominant company is induced by the competitive pressure in the market to pass any cost efficiencies on to the consumer.[291] If a substantial proportion of firms were excluded from the market as a result of such conduct, it would mean that the conduct may be good/efficient on a short-term basis, but on a long-term basis it would harm the market as there would be less and less competition. Long term there would therefore be a risk that the dominant company could raise its prices without fear of competition. The requirement is based on the assumption that efficiencies are best achieved via rivalry between companies.[292] O'Donoghue and Padilla note that there is a contradiction in this requirement in comparison to allowing an efficiency defence. They explain that the efficiency defence is first raised when the company is already found dominant and its conduct to have actual or likely anti-competitive effects. This means that competition is substantially eliminated, yet the conduct may still enhance consumer welfare overall.[293] In other words, the effects of the efficiencies are not given full credit.

It appears that the Commission has merely copied the requirements under Article 81(3) without any particular thought to the consequences of applying these elements to the different issues Article 82 cases bring to the table.[294] However, if the fourth condition works as a trump card which can (always) outweigh any pro-competitive effects, the approach comes into line with current Article 82 case law, but then leaves a question mark as to how the fourth condition can realistically be met.[295] The defence is therefore not particularly helpful for dominant companies and creates legal uncertainty.

---

[289]   Rousseva, Ekaterina, 'Modernising by Eradicating how the Commission's New Approach to Article 81 EC Dispenses with the need to apply Article 82 to Vertical Restraints' (2005) 42 *CMLRev.* 587, p. 633.

[290]   O'Donoghue and Padilla, *supra* n. 18, p. 231.

[291]   The *Article 82 Guidance*, *supra* n. 29, 13.

[292]   Ibid.

[293]   O'Donoghue and Padilla, *supra* n. 18, p. 233.

[294]   For similar opinion see Rousseva (2005) *supra* n. 289, pp. 378 and 382.

[295]   Monti 2007, *supra* n. 43, p. 209.

Rousseva goes one step further than mere duplication or usage of the Article 81(3) exemption under Article 82.[296] Instead, her suggestion is to move the enforcement of tying arrangements together with other vertical restraints to Article 81 only.[297] She defends this suggestion by demonstrating that the recent modernisation of Article 81 has left it in a better position to deal with market power, the only element which in her view has created a segregation of the applied approach to vertical restraints under the two Articles.[298] Although at first this may appear to be a rather radical suggestion, in some ways the suggestion makes sense. First, it would align the EC competition law approach with the US antitrust approach, where tying arrangements are mostly dealt with under Section 1, the equivalent to Article 81. Second, it would create a coherent approach to the same issues: that of vertical restraints. Third, apart from the market power requirement under Article 82 and the requirement of at least two undertakings essential to an Article 81 application, Article 81(3) works as a balancing test, outweighing the pro-competitive effects of an agreement with the anti-competitive effects. This is very similar to the approach taken under Article 82 in respect of anti-competitive effects and objective justifications. The difference is that Article 81(3) has outlined the balancing test in some pre-set conditions. As was shown above, these can work within the Article 82 framework, so why should it be moved?

Moving tying to be dealt with under Article 81 is bound to create some unforeseen problems in relation to the application of Article 81. The US and EC approaches to tying are already extremely similar – moving tying to Article 81 may in fact take the two approaches out of line. Whether tying is dealt with under different legal statutes in the US and the EC is merely a formality in the greater picture. Moreover, dealing with tying purely under Article 81 will at times require the development of an artificial understanding of tying and the relationship between seller and buyer, one which Article 82 need not deal with. Rousseva argues that moving vertical restraints away from Article 82 will allow the competition authorities to focus upon the more serious purely unilateral behaviour.[299] In her definition of purely unilateral behaviour she includes refusal to supply.[300] Nevertheless there is an important overlap between tying arrangements and refusal to supply, and it is therefore not clear-cut that tying necessarily falls under Article 81 all the time or that it is not purely unilateral behaviour. Article 82 case law has clearly indicated that the enforcement does not stop at contractual tying, but includes other forms of

---

[296]   Rousseva (2005) *supra* n. 289.
[297]   Ibid., p. 637.
[298]   Ibid., p. 623.
[299]   Ibid., p. 638.
[300]   Ibid., p. 588.

tying arrangements, such as technological integration. An extensive discussion of this suggestion is beyond the scope of this book. Yet the idea is thought-provoking and raises an interesting question about the relationship between Articles 81 and 82, not least in respect of tying, but also other types of exclusionary abuses.

Although it is a positive development that the Commission is moving towards the approach taken under Article 81(3) and allowing for the anti-competitive effects to be balanced against any efficiencies, the Article 81(3) exemption sets out a pre-fabricated set of conditions, which are, as demonstrated above, not fully appropriate for application under Article 82. Yet it is clear that any balancing test of pro- and anti-competitive effects should allow for a fair share to be passed on to the consumer. The consumer harm test, and to some extent the as-efficient-as competitor test, selected above to assess the anti-competitive effects of the tie will take this into consideration. It is therefore surplus to requirements also to focus upon this in relation to the objective justifications. The replication of the Article 81(3) exemption is therefore rejected.

However, the replication of the Article 81(3) exemption demonstrates that there is a clear imbalance between the standard of proof required for anti-competitive effects and the standard required for objective justifications. The standard for proof of anti-competitive effects is much lower than the one required for objective justification. There is no doubt that a balancing test could be an effective analytical tool; however the standard of proof must be adjusted to ensure that potential efficiencies are equally weighted against potential harms.[301]

7.3. Article 82(d) Exemption – by Nature or According to Commercial Usage

Article 82(d) has a built-in exemption which the EC Courts have so far hindered companies in taking advantage of. The EC Courts' interpretation of commercial usage has rendered it an empty concept, and it therefore offers no protection. A dominant company could find itself in a position where, despite the fact that other companies offered similar bundles, it could be forced to unbundle. The idea is that, in a case such as *Hilti*, if non-dominant companies in the market had also tied their nails with their guns, the tying would firstly, in principle, not be abusive as it would be part of commercial usage and, secondly, it could not be used as an indicator of dominance.

---

[301]   See Kallaugher and Sher for a similar opinion regarding rebates: Kallaugher and Sher, *supra* n. 270, p. 284.

It is debatable how the concept of commercial usage should be interpreted. According to Ahlborn *et al.*, the wording of Article 82(d) means that the absence of commercial usage is a prerequisite for establishing tying.[302] However, the concept of commercial usage could also be an objective justification.[303] Case law shows a remarkably strict interpretation of this concept, an interpretation that leans more towards the latter form – that is an objective justification – rather than its absence-precondition for finding the tying abusive.[304] In *EC Microsoft*, Microsoft naturally attempted to rely on this argument by holding that other operating system vendors also sold operating systems with media players. However, the Commission rejected this argument stating, '[this] argument fails for several reasons ... Microsoft's test looks exclusively to the behaviour of other vendors of the 'tying' product, that is to say, the operating system and disregards that there are independent suppliers of the 'tied' product.'[305] It continued by noting that these other vendors did not tie their specific media players to the operating system but allowed the customers a choice of media player as well as installing only a removable version.[306]

The Commission based its reasoning on the ECJ's judgment in *Tetra Pak II*.[307] In that case the ECJ noted firstly that other companies in the market operated in a different manner from Tetra Pak (a mere 12 per cent of sales were untied[308]).[309] Secondly, it held that even if the dominant company's conduct corresponds to that of commercial usage (that is where *all* firms in the market are following the practice) it still does not preclude the conduct from being found to be abusive when performed by a dominant undertaking.[310] In particular, the latter finding by the ECJ is somewhat of a blow to dominant companies in two respects. Firstly, it fails to clarify the concept of commercial usage. Secondly, it removes much of the basis for legal certainty. The interpretation is so narrow that it leaves little room for manoeuvre for dominant companies, as even normal behaviour can be seen to be abusive once an undertaking becomes dominant. One is left almost inescapably to conclude that commercial usage is in reality an empty concept, and a finding of tying being illegal *per se*.

---

302 Ahlborn, Evans and Padilla (2004) *supra* n. 23, p. 34.
303 Ibid., pp. 34-35.
304 Case C–333/94 *Tetra Pak International SA v Commission (Tetra Pak II)* [1996] ECR I–5951, [1997] 4 CMLR 662, paras 34–35.
305 *EC Microsoft* Commission Decision, para. 822.
306 Ibid., paras. 822–824.
307 Case C–333/94 *Tetra Pak II*, paras. 36–37.
308 Case T–83/91 *Tetra Pak II*, para. 82.
309 Case C–333/94 *Tetra Pak II*, para. 36.
310 Ibid., paras. 36–7.

The CFI's comments on the issue were no more helpful: it merely noted that commercial usage was irrelevant in an industry monopolised by one company.[311] There is not a similar concept or exemption explicitly allowed for under US antitrust law, although the Supreme Court in *Jerrold Electronics*[312] and the Court of Appeals in *US Microsoft III* demonstrated that in US case law there is a willingness to assess market conditions before condemning a tying arrangement. Hence, where the market conditions call for the necessity of tying, either to support the development of a new market or because it is a common practice within a market, US case law will permit this as an objective justification. This also corresponds with the US patent misuse doctrine, where Section 271(d)(5) refers to 'the view of circumstances' and the staple/non-staple distinction of products, which in principle permits the tying of products which by their nature are connected; that is the tied product is specially made to fit the tying (patented) product.

Accepting the findings of Chapter 2, it is clear that tying is a common business practice, and thus legitimising tying makes sense when it is part of the nature of the product or commercial usage in the market.[313] Hence, if a dominant company can prove that the nature of the tied product is specially made to fit the tying product, or it is operating in a market where tying is commercial usage, such as in a high-technology market where speedy innovation and developments of first/second/third generation products and follow-on products are essential to survive, the dominant company should be allowed some leeway in order to ensure continuous innovation and be permitted to tie. Permitting this form of objective justification does not preclude others, such as cost efficiencies and consumer benefits, as case law has allowed for a cluster of defences to be weighted against any anti-competitive effects the tying arrangements may generate.

7.4. Recommending the Article 82(d) Exemption

This section looked at two approaches to objective justifications: the replication of the Article 81(3) exemption under Article 82 and the use of Article 82(d)'s own exemption. Although the Article 81(3) exemption provides for a good alternative to the current objective justification model under Article 82, it appears to overlap somewhat with the anti-competitive effects assessment as well. Given that under section 6 above a balancing test via the consumer harm test was recommended for the anti-competitive effects test, it would be inefficient also to introduce a balancing test for the objective justifications. Instead,

---

[311]   Case T–201/04 *EC Microsoft*, para. 940.
[312]   *Jerrold Electronics* at 557.
[313]   Chapter 2, Sections 6 and 7.

the Article 82(d) exemption is recommended, which is in effect the same exemption that has been allowed for in US case law, and which permits beyond the potential efficiencies to admit the tying arrangement, if the nature of the products or the market conditions dictates that tying is part of commercial usage. This would ensure that innovation is not unnecessarily curbed and that dominant companies can, if the market conditions require, tie.

8. CONCLUSION – SETTING OUT THE NEW REGULATORY MODEL TO TYING

Sections 2 to 7 above have assessed alternatives to the attitude, the format and the individual steps of the current approach to tying under EC competition law in a quest to identify a new and more flexible approach. Each section rejected several alternatives and gave recommendations on the best way forward, all of which are brought together in this section to be combined as the new solution to tying.

Section 2 made it clear that the best solution would be to approach tying from a legal *per se* perspective. Section 3 demonstrated that the current format of a step-test commencing with a set of screens is still a good solution, and this format should thus remain. Sections 4 to 7 assessed each individual step, taking into account the recommendations made in sections 2 and 3. The following three step test is the outcome of that analysis:

(1)   Areeda *et al.*'s consumer demand theory is applied to assess whether the components are two individual products or integrated using the diagram above in section 4; in other words, it is assessed whether there is an separate demand for either of the tied products. The US *Eastman Kodak* definition of consumer demand is applied, meaning that if there is very limited demand, not sufficient to sustain competitors in either market, then it must be concluded that there is no realistic consumer demand for the particular component, and thus the components are likely to be integrated. Moreover, the competition authorities should also look to the market conditions and assess whether there is evidence of the market in general moving towards the integration of products, which would lead to the ceasing of demand for the tied product in the future. Finally, the competition authority may assess whether the product integration creates superior technological efficiencies, which are not achievable from the mere bolting of products and which outweigh any potential anti-competitive effects the technological integration may generate. It is important to include this assessment here rather than under the objective justification section, because it ensures that the product integration is given the full

weight it is worth. If the requirements under 1 are fulfilled the tying arrangement should be permitted. If not the test moves to step 2.

(2)   Importantly, regulation of tying should only commence once the company is dominant; prior to that tying is legal. The market power assessment starts with the table 6.1. Companies holding less than a 40 per cent market share in the tying product market are unlikely to cause harm by the tie. If the company holds between 40 per cent and 60 per cent market share, the tie may cause harm unless the tied product market is sufficiently competitive, or it has low barriers to entry and competitors are able to compete with the dominant company's tying product package to some degree (the as-efficient-as competitor test). If not, or if the company holds more than a 60 per cent market share, a presumption of exclusion of competitors is made and the test moves on to step 3.

(3)   Based on the consumer harm test, an assessment is made of whether the anti-competitive effects of the tying arrangement outweigh any pro-competitive effects. The following elements need to be taken into consideration:

–   is tying part of the nature of the product or commercial usage in the market? If so, the anti-competitive effects must be significant and occurring not to prohibit the tie.
–   Any anti-competitive effects must substantially outweigh the pro-competitive effects for the tying to be harmful. Both can be clusters of effects, which together count towards the weighing. It should be stressed that the standard of proof must be equal for both types of effects; otherwise, the balancing test will never work realistically.

The above steps each carry an equal amount of weight and work together to screen out any pro-competitive ties, but are equally important to identify and catch harmful tying arrangements.

The new test accommodates the findings of Chapter 2, which highlighted the need for a safe haven and the fact that harm was likely to occur only if the company held significant market power in the tying product market and the tied product market had imperfect competition. Yet, the new test follows very much the old approach on both sides of the Atlantic as regards the elements that need to be present; that is two products, market power, anti-competitive effects and defences. This aligns the new test with previous case law and should thus make it easy and attractive for the competition authorities in both jurisdictions to adopt. The new separate product test clearly has its roots in US antitrust law and will thus require the EC competition authorities to take a much more lenient approach to tying than previously, in particular, because it creates an important distinction between contractual tying and technological integration, which should ensure that innovation is not stifled unnecessarily. In comparison, the introduction of the market share cap should cause limited concern for the EC competition authorities, whereas it may be a rather great hurdle for the US competition authorities to overcome, as it is remarkably different from the US case law approach. Although this new test may not be adopted in its entirety, it is hoped that inspiration can and will be drawn from it further to improve and develop the approach to tying.

7. Conclusion

1. CONCLUDING REMARKS

This book has assessed how EC competition law and US antitrust law have approached tying. It identified their shortcomings and inflexibilities, but also highlighted its strengths. Based on the findings of recent economic thinking (in Chapter 2), the EC experience (Chapter 3) and the US experience under both antitrust law (Chapter 4) and IP law (Chapter 5) a new regulatory model was developed (Chapter 6), which particularly considered the recent reform debate of Article 82 and Section 2 and high-technology industries' tendency for innovation through tying and technological integration.

Chapter 2 showed that economic theory dictates that such a simplistic rule such as *per se* illegality, which has been the official approach in US antitrust and the practical consequence under EC competition law, is insufficient to deal with complex tying arrangements such as technological integration, and if innovation is to be promoted and not stifled a more flexible and realistic approach must be sought. To this effect, economists have demonstrated that tying appears to be harmful only when the company undertaking the tying possesses significant market power and the competition in the tied product market is imperfect. Some economists suggested therefore the possibility of introducing certain safe harbours, and importantly argued for a balancing test of the pro- and anti-competitive effects of the tying arrangements. However, such economic requests must be weighed against the legal concerns: the need for legal certainty, time restraints, clear and workable rules, which courts, competition authorities and private parties are reasonably comfortable working with.

EC competition law, and in particular Article 82, contains in fact such a balancing test as recommended by economists, but currently, as was highlighted in Chapter 3, the Commission and the EC Courts have applied an (almost) *per se* illegality approach to tying by placing greater emphasis on the assessment of dominance, instead of the anti-competitive effects and objective justifications assessment creating a disparity between the different elements within the Article 82 analysis. The result is that once a company is found dominant, the type of behaviour it is engaged in becomes central to whether it is found to infringe Article 82, rather than whether the behaviour is causing any significant harm to competition. This approach is referred to as a form-based approach and characterised by a focus on the conduct and the application of an assumption of (likely future) anti-competitive effects. Consequently, any (potential) anti-competitive effects will always triumph over any objective justifications. It was found in Chapter 3 that the current application of the balancing test under Article 82 is defective and a re-balancing of this test as a minimum is essential to align the approach to tying under Article 82 with modern day economic thinking and market development.

Chapter 4 looked upon the US antitrust approach to tying and found that it shared many common characteristics with the EC competition law approach. In particular, both approaches apply the same step-test,[1] and includes similar elements, such as the consumer demand test to identify whether there are two products, the establishment of market power, an assessment of the effects upon the tied product market and an equally high threshold for any objective justifications raised for the tie. The US antitrust approach however differed in two remarkable aspects. First, US antitrust case law has created a clear distinction in the treatment of contractual tying and technological integration to the point where the latter receives more favourable treatment under the assumption that this will benefit innovation. The second aspect is the attitude to tying in general, which has been demonstrated to be a more flexible approach, one that in particular over the last three decades (commencing with *Jefferson Parish*[2]) has developed to a point where it is no longer a strict *per se* illegality approach which is applied, but a truncated *rule of reason*, which will consider the facts of the individual cases to ensure that if there are specific circumstances present, such as the development of a new market,[3] or the industry in itself leans towards innovation via integration,[4] these will be given sufficient weight in the Court's decision whether the tie is illegal or not. This approach demonstrates an attitude more in line with recent economic thinking in comparison to EC competition law, and the *Section 2 Report* from the DoJ offers considerable hope that this is the line which will be toed in the future.[5] Yet the US courts seem reluctant to shake off the *per se* illegality label attached to tying arrangements under US antitrust law. Therefore, although US antitrust law has

---

[1]   See Chapter 6, section 3.
[2]   *Jefferson Parish Hospital District No. 2 v Hyde*, 466 U.S. 2, 104 S Ct. 1551, 80 L.Ed. 2d2 (1984).
[3]   *United States v Jerrold Electronics Corp.*, 187 F. Supp. 545 (E.D.Pa. 1960), *affirmed Jerrold Electronics Corp. v United States*, 365 U.S. 567, 81 S. Ct. 755, 5 L.Ed. 2d 806 (1961).
[4]   See *US v Microsoft Corp.*, 253 f.3d 34 (D.C. Cir. 2001) (*US Microsoft III*).
[5]   US Department of Justice, 'Competition and Monopoly: Single-Firm Conduct Under Section 2 of the Sherman Act (2008)' (*Section 2 Report*) (www.usdoj.gov/atr/public/reports/ 236681.htm).

developed further towards a more economic and flexible approach to tying, it has not moved far enough.

Equally, Chapter 5 showed that under US IP law the Congress has stepped in to ensure that tying arrangements of patented products can take place unless the patent owner holds market power. This *per se* legality approach illustrates a positive outlook upon tying arrangements and indicates that tying arrangement in certain circumstance can facilitate innovation.

The *per se* legality approach created the framework surrounding the new regulatory model suggested for tying in Chapter 6. The idea was to create a model approach which not only is in line with economic thinking, but also provides a set of workable rules for private parties, competition authorities and courts. The model contains four vital steps to ensure just that: primarily two forms of safe harbours which can weed out pro-competitive tying arrangements, one relying on the necessity of two separate products, the other on significant market power. As pointed out by economists, the latter is a core element in establishing whether a tie will be harmful to competition. Therefore, above a certain market power threshold a company will be assumed to be excluding competitors from the market because of its tie. Where a company holds less market power an assessment of whether equally efficient competitors are excluded from the market will be made based on barriers to entry, the ability to offer matching product packages, and the level of competition in the tied product market. If a company is found to exclude competitors a further assessment of additional anti-competitive effects will be made and these will be balanced against any objective justifications, such as and in particular the nature of the product and the commercial usage within the industry. The test combines the economic thinking that tying is often pro-competitive and beneficial for competition with the form-based and traditional *per se* illegality approach seen in case law. It does not offer a full-on green light to tying arrangements under a presumption that all tying arrangements are pro-competitive and should be treated under a *rule of reason* approach. Instead, the test maintains the format of the traditional approach to tying under both US and EC competition law – the step test, which means that tying will remain regulated. However, the individual steps in the new proposed test are the ones that create the greater flexibility. For instance, the expanded consumer demand test and the dominance assessment distinguish between harmful and pro-competitive ties under a clear set of rules.

The expanded consumer demand test is more sophisticated than the standard consumer demand test currently applied because it assesses not just the consumer demand for the tied product, but also that of the tying product, their relationship and the specific market conditions surrounding the tie. At the same time it permits a distinction between contractual tying and technological integration which has so far not been available under EC competition law. The dominance assessment includes an initial anti-competitive effects screening with the inclusion of the as-efficient-as competitor test, and places greater emphasis upon the anti-competitive effects of the conduct rather than the dominance of the company in question. The test therefore transfers the emphasis back onto the harm caused by the behaviour rather than the current approach under Article 82 (less so for the Sherman Act), which is much more focused upon the establishment of dominance.

The suggested new approach sits well with some aspects of the discussed reform of Article 82 and Section 2; in particular, it offers a more economic approach but maintains clear and workable rules. The proposed new approach establishes an alternative to the current test for tying, which is both more economic and flexible in its application, which will protect technological integration and thereby a special form of innovation, but is also aligned with the traditional approach by creating clear rules and legal certainty.

2. TECHNOLOGICAL INTEGRATION

In the US, the courts early on effectively carved out a safe haven for technological integrated products, thus creating an important discrepancy between technological integration and contractual tying. In comparison, in Europe, the technological integration discussion has so far been avoided by having a very high standard of proof of when a product is no longer a product, but a mere component of another.[6]

EC Microsoft raised concerns whether the high standard of proof is appropriate considering that technological integration is conduct which is much more delicate and has more severe consequences than finding a contractual tying arrangement illegal, such as stifling innovation. At the same time, it should be questioned whether the courts are best suited to assess whether a technological integration is more efficient than the purchase of two components separately.[7] Some US courts' motivation for creating such a distinction may not have been purely for the benefit of innovation, but also a reluctance to engage in a technical assessment of the tying arrangement. Therefore, although a test of technological integration cannot avoid the technical aspects, other elements such as market conditions and market participants' perception of trends and product development within the market need to be included to alleviate the courts and competition authorities of the burden of deciding upon the right technology.

---

[6] Case T–201/04 *EC Microsoft*, para. 918.
[7] See Spulber, D., 'Unlocking Technology: Antitrust and Innovation' (2008) 4 *Journal of Competition Law and Economics* 915, pp. 915–916 and 956.

The current consumer demand test applied on both sides of the Atlantic to identify whether there are two separate products has been shown to be too simplistic and restrictive to provide adequate protection, or more importantly ensure that innovation and product development are not stifled. The main problem with the test is that the courts have been too focused upon the requirement of independent producers of the tied product, but have given no or limited consideration to market conditions, (the lack of) the existence of a separate demand for the tying product, and the net efficiencies generated from novel product integration. The high standard of proof adopted under EC competition law further hinders this by not sufficiently taking into account the change in consumer perception of products in innovative and fast-moving industries. These points all lead to the conclusion that there is a need in the US antitrust case to preserve the distinction between contractual tying and technological integration and in the EC competition law case create such a division to ensure that innovation and product development are not held back unnecessarily. This book has suggested a test which will permit a safe haven for technological integration. The test analyses the tying arrangement from three different perspectives: the consumer perspective by assessing the demand for both products, the supply/manufacturer perspective by examining the market conditions and trends within the market, and finally, an objective perspective, which considers whether the tying arrangement offers superior efficiencies which cannot be achieved by the mere bolting of the products together and which substantially outweigh any anti-competitive effects the tying arrangement may generate. If all conditions are fulfilled the tying arrangement should be seen as technological integration and saved from further intervention from the competition authorities. Creating this safe haven should ensure that innovation is not stifled unreasonably, but at the same time that the competitive process is also not obstructed.

3.   FUTURE CASES

The Commission has already continued its crusade against Microsoft and issued it with a statement of objection for, amongst other things, the tying of Internet Explorer with Windows. This new investigation is based on complaints made by competitors to Microsoft over the last two years and targets some of Microsoft's core products.[8] The facts of the case appear to be

---

8   Waters, Richard, 'EU launches new Microsoft antitrust probe', *Financial Times*, 14 January, 2009 (www.ft.com/cms/s/0/2ce90532-c2c1-11dc-b617-0000779fd2ac,_i_email=y.html).

very similar to those of the previous *EC Microsoft* case, just another product – in fact the one that was the centre of attention in the *US Microsoft* cases. Microsoft is accused of shielding Internet Explorer from competition by tying it to Windows, providing it with a distribution advantage and an artificial incentive for software developers and content providers to write specifically for Internet Explorer rather than competing products.[9] The Commission has now successfully brought one case against Microsoft and thus knows which buttons to push. It is therefore doubtful that the Commission will change tactics and show a more lenient approach to technological integration in comparison to previous cases. It is also uncertain whether Microsoft has a chance at all to defend its actions – or perhaps it has learned from its previous mistakes and will this time round offer sufficient evidence of 'superior technical product performance' generated from the tie. In a BBC programme about Bill Gates,[10] the management of Microsoft gave the impression that Microsoft was keen to cooperate with the competition authorities, but reluctant to change certain business strategies, as they saw nothing wrong with ferocious competition. The question is therefore how many beatings can/will Microsoft take from the competition authorities before it changes it business strategies?

In the US there has been a change in government. Last time, there was a change in government Microsoft was incidentally saved from being divided into two companies.[11] However, as President Barack Obama takes his seat in the Oval Office the question remains whether this will lead to increased concern about Microsoft's behaviour. The President's selected antitrust enforcement team is certainly expected to be tougher than Bush's, especially on mergers, but whether this transfers into unilateral conduct remains to be seen[12]. One thing that is sure is that Microsoft continues to be monitored in relation to its compliance with the final Microsoft judgments.[13]

It is clear that the *Microsoft* cases on both sides of the Atlantic have taught us a great deal about tying, and in particular technological integration.

---

[9]   Keizer, Gregg, 'Update: EU hits Microsoft with new antitrust charges', *Computerworld*,   16   January,   2009   (www.computerworld.com/action/article.do?command=viewArticleBasic&articleId=9126221).
[10]   BBC 2, 'Money Programme: Bill Gates: How a Geek Changed the World', broadcast 20 June 2008.
[11]   Ibid. and Evans, David, Nicholas, Albert, and Schmalensee, Richard, 'United States v Microsoft: Did Consumers Win?'(2005) 1 *Journal of Competition Law and Economics* 497, pp. 507–508.
[12]   The withdrawal of the Section 2 Report certainly seems to confirm this stricter attitude, see press release 09-459 'Justice Department Withdraws Reports on Antitrust Monopoly Law' 11 May 2009 (www.usdoj.gov/opa/pr/2009/May/09-at459.html).
[13]   US Department of Justice, Antitrust Division (http://www.usdoj.gov/atr/cases/ms_index.htm).

Undoubtedly, these cases were of an extreme nature, but the US courts have demonstrated a great willingness to ensure that the legal approach to tying moves with time and economic understanding. In the EU, a more rigid approach to tying continues, and with it the risk of stifling innovation and product development. However, these concerns have been somewhat over-shadowed by the greater publicity battle between the Commission and Microsoft, and this severely limits the chance of the Commission adopting a more economic approach towards tying and creating the all-important distinction between contractual tying and technological integration to encourage and promote innovation.

Bibliography

BOOKS

Amato, Giuliano, *The Antitrust and the Bounds of Power, the Dilemma of Liberal Democracy in the History of the Market* (Hart Publishing, Oxford, 1997).

Anderman, Steve, *EC Competition Law and Intellectual Property Rights, the Regulation of Innovation* (Oxford University Press, Oxford, 1998).

Anderman, Steve and Kallaugher, John, *Technology Transfer and the New EU Competition Rules, Intellectual Property Licensing after Modernisation* (Oxford University Press, Oxford, 2006).

Areeda, Philip., Elhauge, Einer, and Hovenkamp, Herbert, *Antitrust Law, An Analysis of Antitrust Principles and their Application* (2nd edn, Aspen Publishers, New York, 2004), vol. X.

Areeda, Philip, Kaplow, Louis and Edlin, Aaron, *Antitrust Analysis, Problems, Text, and Cases* (6th edn, Aspen Publishing, New York, 2004).

Bishop, Simon and Walker, Mike, *The Economics of EC Competition Law* (Sweet and Maxwell, London, 1999).

Bork, Robert H., *The Antitrust Paradox: A Policy at War with Itself* (The Free Press, New York, 1993).

Cornish, William R. and Llewelyn, David, *Intellectual Property: Patents, Copyright, Trade Marks and Allied Rights* (5th edn, Sweet and Maxwell, London, 2003).

Faull, Jonathan and Nikpay, Ali (eds), *The EC Law of Competition* (Oxford University Press, Oxford, 2007).

Fejø, Jens, *EU-Konkurrenceret* (2nd edn, Jurist- og Økonomforbundets Forlag, Copenhagen, 1997).

Gellhorn, Ernest, Kovacic, William E. and Calkins, Stephen, *Antitrust Law and Economics in a Nutshell* (5th edn, Thomsen West, St. Paul, Minn., 2004).

Gerber, David, *Law and Competition in the Twentieth Century Europe: Protecting Prometheus* (Oxford University Press, Oxford, 1998).

Glader, Marcus, *Innovation Markets and Competition Analysis, EU Competition Law and US Antitrust Law* (Edward Elgar, Cheltenham, 2006).

Govaere, Inge, *The Use and Abuse of Intellectual Property Rights in E.C. Law* (Sweet and Maxwell, London, 1996).

Groves, Peter, *Intellectual Property with Competition Law and Practice* (London Guildhall-Cavendish, London, 1994).

Holyoak, Jon and Torremans, Paul, *Intellectual Property Law* (2nd edn, Butterworths, London, 1998).

Hovenkamp, Herbert, Janis, Mark D., and Lemley, Mark A., *IP and Antitrust, An Analysis of Antitrust Principles Applied to Intellectual Property Law* (Aspen Law & Business, New York, 2004).

Hylton, Keith N., *Antitrust Law, Economic Theory & Common Law Evolution* (Cambridge University Press, Cambridge, 2003).

Jones, Alison and Sufrin, Brenda, *EC Competition Law, Text, Cases and Materials* (3rd edn, Oxford University Press, Oxford, 2007).

Korah, Valentine, *Intellectual Property Rights and the EC Competition Rules* (Hart Publishing, Oxford and Portland, Oregon, 2006).

Langer, Jurian, *Tying and Bundling as a Leveraging Concern under EC Competition Law* (Kluwer Law International, Alphen aan den Rijn, 2007).

Martin, Stephen, *Industrial Economics, Economic Analysis and Public Policy* (2nd edn, Prentice Hall, Englewood Cliffs, NJ, 1993).

Monti, Giorgio, *EC Competition Law* (Cambridge University Press, Cambridge, 2007).

Motta, Massimo, *Competition Policy – Theory and Practice* (Cambridge University Press, New York, 2004).

Neale, Alan D. and Goyder, Dan G., *The Antitrust Laws of the United States of America, A Study of Competition Enforced by Law* (3rd edn, Cambridge University Press, Cambridge, UK and New York, 1980).

O'Donoghue, Robert and Padilla, A. Jorge, *The Law and Economics of Article 82 EC* (Hart Publishing, Oxford and Portland, Oregon, 2006).

Peritz, Rudolph, *Competition Policy In America, History, Rhetoric, Law* (Oxford University Press, Oxford, 2001).

Posner, Richard A., *Antitrust Law: An Economic Perspective*, University of Chicago Press, Chicago, Ill., 1976).

Rahnasto, Ilkka, *Intellectual Property Rights, External Effects and Antitrust Law, Levering IPRs in the Communications Industry* (Oxford University Press, Oxford, 2003).

Schechter, Roger and Thomas, John R., *Intellectual Property, the Law of Copyrights, Patents and Trademarks* (Thomson/West, St. Paul, Minn., 2003).

Sloman, John, *Economics* (5th edn, Financial Times Prentice Hall (Pearson Education), Harlow, 2003).

Stigler, George, *The Organization of Industry* (Richard D Irwin, Inc. Homewood, Ill., 1968).

Sullivan, Lawrence A. and Grimes, Warren S., *The Law of Antitrust: An Integrated Handbook* (West Group, St Paul, Minn., 2000).

Tirole, Jean, *The Theory of Industrial Organisation* (MIT Press, Cambridge, Mass., 1978) (reprinted 2003).

Van den Bergh, Roger J. and Camesasca, Peter D., *European Competition Law and Economics: A Comparative Perspective* (2nd edn, Thomson Sweet & Maxwell, London, 2006).

CHAPTERS IN BOOKS

Coch, Nicholas and Chen, Heidi, 'Specific Practices that have been Challenged as Misuse', chapter 2 in ABA Section of Antitrust Law, *Intellectual Property Misuse; Licensing and Litigation* (ABA, Chicago, 2000).

Fisher, Franklin M., 'Innovation and Monopoly Leveraging' in Ellig, Jerry (ed.), *Dynamic Competition and Public Policy, Technology, Innovation and Antitrust Issues* (Cambridge University Press, New York, 2001).

Gordon, George and Hoerner J. Robert, 'Overview and Historical Development of the Misuse Doctrine', chapter I in ABA Section of Antitrust Law, *Intellectual Property Misuse; Licensing and Litigation* (ABA, Chicago, 2000).

Langer, Jurian, 'Bundling, A Four-Step Test to Assess the Exclusionary Effects of Bundling under Article 82 EC', Chapter 8 in Amato, Giuliano and Ehlermann, Claus-Dieter (eds), *EC Competition Law, A Critical Assessment* (Hart Publishing, Oxford, 2007).

Monti, Giorgio, 'Article 82 EC and New Economy Markets', Chapter 2 in Graham, Cosmo and Smith, Fiona (eds), *Competition, Regulation and the New Economy* (Hart Publishing, Oxford, 2004).

Rousseva, Ekaterina, 'Abuse of Dominant Position Defences, Objective Justification and Article 82 EC in the Era of Modernisation', Chapter 10 in Amato, Giuliano and Ehlermann, Claus-Dieter (eds), *EC Competition Law, A Critical Assessment* (Hart Publishing, Oxford, 2007).

Vickers, John, 'Abuse of Market Power', chapter 11 in Buccirossi, Paolo (ed.), *Handbook of Antitrust Economics* (MIT Press, Cambridge, Mass., 2008).

ARTICLES

Adams, William James and Yellen, Janet L., 'Commodity Bundling and the Burden of Monopoly' (1976) 90 *The Quarterly Journal of Economics* 475–498.

Ahlborn, Christian, Evans, David S. and Padilla, Atilano Jorge, 'Competition Policy in the New Economy: Is European Competition Law Up to the Challenge?' (2001) *European Competition Law Review* 22(5), 156–167.

Ahlborn, Christian, Evans, David S., and Padilla, A. Jorge, 'The Antitrust Economics of Tying: A Farewell to *Per Se* Illegality' (2004) *Antitrust Bulletin*, 287–341.

Ahlborn, Christian, Denicolò, Vincenzo, Geradin, Damien and Padilla, A. Jorge, 'DG Comp's Discussion Paper on Article 82: Implications of the Proposed Framework and Antitrust Rules for Dynamically Competitive Industries', 31st March 2006 (http://papers.ssrn.com/sol3/papers.cfm?abstract_id=894466).

Al-Dabbah, Maher M., 'Conduct, Dominance and Abuse in "Market Relationship": Analysis of Some Conceptual Issues under Article 82 EC' (2000) 21 *European Competition Law Review* 45–50.

Andrews, P., 'Aftermarket Power in the Computer Service Market: The Digital Undertaking' (1998) 19 *European Competition Law Review* 176–181.

Anthony, Sheila F., 'Antitrust and Intellectual Property Law: From Adversaries to Partners' (2000) 28 *AIPLA Quarterly Journal* 1 (www.ftc.gov/speeches/other/aipla.htm).

Apon, J., 'Cases Against Microsoft: Similar Cases, Different Remedies' (2007) 28 *European Competition Law Review* 327–336.

Appeldoorn, Jochen, 'He Who Spareth his Rod, Hateth His Son? Microsoft, Super-dominance and Article 82 EC' (2005) *European Competition Law Review* 653–658.

Art, Jean-Yves and McCurdy, Gregory, 'The European Commission's Media Player Remedy in its Microsoft Decision: Compulsory Code Removal Despite the Absence of Tying or Foreclosure' (2004) 25 *European Competition Law Review* 694–707.

Azevedo, Joao Pearce de, and Walker, Mike, 'Market Dominance: Measurement Problems and Mistakes' (2003) 24 *European Competition Law Review* 640–643.

Bakos, Yannis and Brynjolfsson, Erik, 'Bundling Information Goods: Pricing, Profits and Efficiency' (1999) 45 *Management Sci.* 1613.

Baldwin, William D. and McFarland, David, 'Tying Arrangements in Law and Economics' (1963) 8 *Antitrust Bulletin* 743–80.

Bavasso, Antonio, 'The Role of Intent under Article 82 EC: From "Flushing the Turkeys" to "Spotting Lionesses in Regents Park" ' (2005) 26 *European Competition Law Review* 616–623.

Bowman, Ward S., 'Tying Arrangements and the Leverage Problem' (1957) 67 *The Yale Law Journal* 19–36.

Burchfiel, Kenneth J., 'Patent Misuse and Antitrust Reform: "Blessed Be the Tie?" ' (1991) 4 *Harvard Journal of Law and Technology* 1–108.

Calkins, Stephen, 'Patent Law: The Impact of the 1988 Patent Misuse Reform Act and Noerr–Pennington Doctrine on Misuse Defenses and Antitrust Counterclaims' (1989) 38 *Drake Law Review* 175–228.

Carbajo, Jose, De Meza, David and Seidman, Daniel J., 'A Strategic Motivation for Commodity Bundling' (1990) 38 *Journal of Industrial Economics* 283–298.

Carlton, Dennis W. and Waldman, Michael, 'The Strategic Use of Tying to Preserve and Create Market Power in Evolving Industries' (2002) 33 *RAND Journal of Economics* 194–220.

Chen, Yongmin, 'Equilibrium Product Bundling' (1997) 70 *Journal of Business* 85–103.

Choi, Jay Pil, 'Tying and Innovation: A Dynamic Analysis of Tying Arrangements', revised May 2002 (www.msu.edu/~choijay/Tying.pdf (9 July 2004).

Choi, Jay Pil and Stefanadis, Christodoulos, 'Tying, Investment, and the Dynamic Leverage Theory' (2001) 32 *RAND Journal of Economics* 52–71.

Dhar, Ravi and Nowles, S.M., 'To Buy or Not to Buy: Response Mode Effects on Consumer Choice' (2004) 41 *Journal of Marketing Research* 423–432.

Diaz, F.E. Gonzalez and Garcia, A.L., 'Tying and Bundling under EU Competition Law: Future Prospects' (2007) 3 *Competition Law International* 13–19.

Dolmans, Maurits and Graf, Thomas, 'Analysis of Tying Under Article 82 EC: The European Commission's Microsoft Decision in Perspective' (2004) 27 *World Competition* 225–244.

Dolmans, Maurits and Piilola, Anu, 'The Proposed New Technology Transfer Block Exemption, Is Europe Really Better Off than with the Current Regulation?' (2003) 26 *World Competition* 541–565.

Eilmansberger, Thomas, 'How to Distinguish Good from Bad Competition Under Article 82 EC: In Search of Clearer and more Coherent Standards for Anti-competitive Abuses' (2005) 42 *Common Market Law Review* 129–177.

Evans, David S., 'All the Facts That Fit: Square Pegs and Round Holes in the US v. Microsoft' 22 *Regulation* 1999, (1–10). (http://www.cato.org/pubs/regulation/regv22n4/evans.pdf)

Evans, David S., Nichols, Albert L. and Schmalensee, Richard, 'United States v Microsoft: Did Consumers Win?' (2000) 1 *Journal of Competition Law and Economics* 497–538.

Evans, David S. and Padilla, A.Jorge, 'Tying Under Article 82EC and the Microsoft Decision: A Comment of Dolmans and Graf' [2004] 27 *World Competition: Law and Economics Review* 503–512.

Evans, David S., Padilla, A.Jorge and Polo, Michele, 'Tying in Platform Software: Reasons for a *Rule-of-Reason* Standard in European Competition Law' (2002) 25 *World Competition* 509–514.

Evans, David S, and Salinger, Michael, 'Why Do Firms Bundle and Tie? Evidence from Competitive Markets and Implications for Tying Law' (2005) 22 *Yale Journal on Regulation* 37–89.

Feldman, Robin Cooper, 'Defensive Leveraging in Antitrust' (1999) 87 *Georgetown Law Journal* 2079–2115.

Furse, Mark, 'On a Darkling Plain: The Confused Alarms of Article 82 EC' (2004) 25 *European Competition Law Review* 317–319.

Gastle, Charles and Boughs, Susan, ' Microsoft III and the Metes and Bounds of Software Design and Technological Tying Doctrine' (2001) 6 *Virginia Journal of Law and Technology*, 7, 7–120.

Gilbert, Richard J. and Katz, Michael L., 'An Economist's Guide to *U.S. v. Microsoft*' (2001) 15 *Journal of Economic Perspectives* 25–44.

Glazer, Kenneth L. and Lipsky, Abbott B., 'Unilateral Refusals to Deal under Section 2 of the Sherman Act' (1995) 63 *Antitrust Law Journal* 749–800.

Harbord, David and Hoehn, Tom, 'Barriers to Entry and Exit in European Competition Policy' (1994) 14 *International Review of Law and Economics* 411–433.

Harchuck, Kara E., 'Microsoft IV: The Dangers to Innovation posed by the Irresponsible Application of a Rule of Reason Analysis to Product Design Claims' (2002) 97 *Northwestern University Law Review* 395–438.

Hawker, Norman, 'Consistently Wrong: The Single Product Issue and the Tying Claims Against Microsoft' (1998) 35 *California Western Law Review* 1–39.

Heal, Madeleine, 'Loosening the Ties: Tie-in Clauses to be Assessed under "Effects"-based Competition Act 1998' (1999) 21 *European Intellectual Property Review* 414–416.

Hovenkamp, Herbert, 'IP Ties and Microsoft's Rule of Reason' [2002] *The Antitrust Bulletin* 369–422.

Hylton, Keith N. and Salinger, Michael, 'Tying Law and Policy: A Decision-theoretic Approach' (2001) 69 *Antitrust Law Journal* 469–526.

Jacobs, Michael, 'Third Line Forcing and Tying Arrangements – Some Comments on the United States' Position', article published on Blake Dawson Waldron web site: Competition and Consumer Protection (www.bdw.com.au/areas/tradespractices/tpa-new-4.htm).

Jacobson, Jonathan M. and Sher, Scott A., ' "No economic sense" Makes No Sense for Exclusive Dealing' (2006) 73 *Antitrust Law Journal* 779–801.

Kallaugher, Jon and Sher, Brian, 'Rebates Revisited: Anti-competitive Effects and Exclusionary Abuse under Article 82' (2004) 25 *European Competition Law Review* 263–285.

Kaplow, Louis, 'Extension of Monopoly Power through Leverage' (1985) 85 *Columbia Law Review* 515–555.

Katz, Michael L. and Shapiro, Carl, 'Network Externalities, Competition, and Compatibility' (1985) 75 *American Economic Review* 424–440.

Katz, Michael L. and Shapiro, Carl, 'Systems Competition and Network Effects' (1994) 8 *Journal of Economic Perspectives* 93–115.

Korah, Valentine, 'The Interface Between Intellectual Property and Antitrust: The European Experience' (2002) 69 *Antitrust Law Journal* 801–839.

Krim, Jonathan, 'EU Orders Microsoft to Modify Windows', *Washington Post*, 23 December 2004, AOI.

Kühn, Kai-Uwe, Stillman, Robert and Caffarra, Christina, 'Economic Theories of Bundling and their Policy Implications in Abuse Cases: An Assessment in Light of the Microsoft Case' (2005) 1 *European Competition Journal* 85–121.

Laddie, Sir Hugh, 'National IP Rights: A Moribund Anachronism in a Federal Europe' (2001) 23 *European Intellectual Property Review* 402–408.

Lam, David K., 'Revisiting the Separate Products Issue' (1999) 108 *Yale Law Journal* 1441–4148.

Lang, John Temple, 'Monopolisation and the Definition of "Abuse" of a Dominant Position under Article 86 EEC Treaty' (1979) 16 *CMLRev.* 345.

Lawsky, David, 'EU Set to Rule Against Microsoft' Reuters, 23 March 2004 (www.reuters.com).

Le, Net, 'What does "Capable of Eliminating All Competition" Mean' (2005) 26 *European Competition Law Review* 6–10.

Lehmann, Michael, 'The Theory of Property Rights and the Protection of Intellectual and Industrial Property' (1985) 16 *International Review of Industrial Property and Copyright Law* 525, p. 531–32.

Lemley, Mark, 'The Economic Irrationality of the Patent Misuse Doctrine' (1990) 78 *California Law Review* 1599–1631.

Lind, Robert C. and Muysert, Paul, 'Innovation and Competition Policy, Challenges for the New Millennium' (2003) 24 *European Competition Law Review* 87–92.

Meese, Alan J., 'Monopoly Bundling in Cyberspace: How Many Products does Microsoft Sell?' [1999] 44 *The Antitrust Bulletin* 65-116.

Melamed, A. Douglas, 'Exclusionary Conduct Under the Antitrust Laws: Balancing, Sacrifice, and Refusal to Deal' (2005) 20 *Berkeley Tech. Law Journal* 1247–1267.

Merges, Robert P., 'Reflections on Current Legislation Affecting Patent Misuse' (1988) 70 *Journal of the Patent Office Society* 793–804.

Montgomery, W., 'The Presumption of Economic Power for Patented and Copyrighted Products in Tying Arrangements' (1985) 85 *Columbia Law Review* 1140–1156.

Monti, Giorgio, 'The Concept of Dominance in Article 82' [2006] *European Competition Journal* 31–52.

Ong, Burton, 'Building Brick Barricades and Other Barriers to Entry: Abusing a Dominant Position by Refusing to Licence Intellectual Property Rights' (2005) 26 *European Competition Law Review* 215–224.

Page, William H. and Lopatka, John E., 'The Dubious Search for "Integration" in the *Microsoft* Trial' (1999) 31 *Connecticut Law Review* 1251–1274.

Peritz, Rudolph, 'Theory and Fact in Antitrust Doctrine: Summary Judgment Standards, single-brand aftermarkets and the clash of microeconomic models' (2000) 45 *Antitrust Bulletin* 887–920.

Pitofsky, Robert, 'Challenges of the New Economy: Issues at the Intersection of Antitrust and Intellectual Property' (2001) 69 *Antitrust Law Journal* 913–924.

Price, Diana R., 'Abuse of a Dominant Position – The Tale of Nails, Milk Cartons and TV Guides' (1990) 11 *European Competition Law Review* 80–90.

Ridyard, Derek, 'Tying and Bundling – Cause for Complaint?' (2005) 26 *European Competition Law Review* 316–319.

Ritter, Cyril, 'Refusal to Deal and 'Essential Facilities': Does Intellectual Property Require Special Deference Compared to Tangible Property?' (2005) 28 *World Competition: Law and Economics Review* 281–298. (http://papers.ssrn.com/sol3/papers.cfm?abstract_id=726683).

Rousseva, Ekaterina, 'Modernising by Eradicating how the Commission's New Approach to Article 81 EC Dispenses with the Need to Apply Article 82 to Vertical Restraints' (2005) 42 *CMLRev.* 587–638.

Rushe, Dominic, 'Microsoft Braced for Big Fines by EU', *The Sunday Times*, 21 March 2004.

Salinger, Michael A., 'Graphical Analysis of Bundling'(1995) 68 *The Journal of Business* 85–98.

Salop, Steven C., 'Exclusionary Conduct, Effect on Consumers, and the Flawed Profit-Sacrifice Standard' (2006) 73 *Antitrust Law Journal* 311–374.

Sinclair, Duncan, 'Abuse of Dominance at a Crossroads – Potential Effect, Object and Appreciability under Article 82 EC' (2004) 25 *European Competition Law Review* 491–501.

Spulber, Daniel, 'Unlocking Technology: Antitrust and Innovation' (2008) 4 *Journal of Competition Law and Economics* 915–966.

Tirole, Jean, 'The Analysis of Tying Cases: A Primer' (2005) 1 *Competition Policy International* 1.

Tom, Willard K. and Newberg, Joshua A., 'Antitrust and Intellectual Property: From Separate Spheres to Unified Field' (1997) 66 *Antitrust Law Journal* 167–229.

'Use of Tie-Ins in New Industries, the Notes and Comments' (1961) 70 *Yale Journal of Law* 804–811.

Veljanovski, Cento, 'EC Antitrust in the New Economy: Is the European EC Commission's View of the Network Economy Right?' (2001) 22 *European Competition Law Review* 115–121.

Waelbroek, Denis, 'The Compatility of Tying Agreements with Antitrust Rules: A Comparative Study of American and European Rules' [1987] *Yearbook of European Law* 39–57.

Webb, Jere M. and Locke, Lawrence A., 'Intellectual Property Misuse: Developments in the Misuse Doctrine' (1999) 4 *Harvard Journal of Law and Technology* 257–267.

Werden, Gregory J., 'Identifying Exclusionary Conduct under Section 2: The "No Economic Sense" Test' (2006) 73 *Antitrust Law Journal* 413–433.

Whinston, Michael D., 'Exclusivity and Tying in *U.S. v. Microsoft:* What We Know, and Don't Know' (2001) 15 *Journal of Economic Perspectives* 63–80.

Whinston, Michael D., 'Tying, Foreclosure, and Exclusion' (1990) 80 *The American Economic Review* 837–859.

Williams, Mark, 'Sega, Nintendo and Aftermarket Power: The Monopolies and Mergers Commission Report on Video Games' (1995) 16(5) *European Competition Law Review* 310–318.

Woodrow De Vries, Michael, 'United States v Microsoft' (1999) 14 *Berkeley Technology Law Journal* 303–322.

OTHER (WEBSITES/SPEECHES/REPORTS)

BBC 2, 'Money Programme: Bill Gates: How a Geek Changed the World', broadcast 20 June 2008.

Bellis, Jean-François, 'The Microsoft Judgment', Conference, 25 September 2007, BIICL, London.

Charles River Associates, 'Innovation and Competition Policy, Part I – Conceptual Issues', Economic Discussion Paper 3, March 2002, Report prepared for the Office of Fair Trading by Charles River Associates, (www.oft.gov.uk/News/Publications).

Delrahim, Mekan and Murphy, Frances, comments given at 'The Microsoft Judgment' Conference, 25 September 2007, BIICL, London.

Drexl, Josef, 'Refusal to License and IP Right', – Comments by the Max Planck Institute for Intellectual Property and Tax Law, Munich (http://ec.europa.eu/comm/competition/antitrust/others/drexl.pdf).

Economic Advisory Group on Competition Policy (EAGCP), 'An Economic Approach to Article 82', July 2005 (http://ec. europa.eu/comm/competition/publications/studies/eagcp_july_21_05.pdf).

Forrester, Ian S., 'Compulsory Licensing in Europe: a Rare Cure to Aberrant National Intellectual Property Rights?' presentation at 'Competition and Intellectual Property Law and Policy in the Knowledge-Based Economy: Comparative Law Topics', Department of Justice/Federal Trade Commission Hearings, Department of Justice, Great Hall, Washington, DC, 22 May 2002.

Fromm, Jeffrey B. and Skitol, Robert A., 'Harmonisation of the IP Misuse

Doctrine and Antitrust Law: A Call for Help from the Agencies and Congress' theantitrustsource, January 2003 (www.antitrustsource.com (http://www.abanet.org/antitrust/source/01-03/frommskitol.pdf).

Hellström, Per, 'The Microsoft Judgment', Conference, 25 September 2007, BIICL, London.

Keizer, Gregg, 'Update: EU hits Microsoft with new antitrust charges', *Computerworld*, 16 January 2009 (www.computerworld.com/action/article.do?command=viewArticleBasic&articleId=9126221).

Kroes, Neelie, 'Preliminary Thoughts on Policy Review of Article 82', Speech at the Fordham Corporate Law Institute, New York, 23 September 2005 (http://europa.eu/rapid/pressReleasesAction.do?reference=SPEECH/05/53 7&format=HTML&aged=0&language=EN&guiLanguage=en).

Marsden, Philip, comments given at 'The Microsoft Judgment' Conference, 25 September 2007, BIICL, London.

McDonald, J. Bruce, 'The Struggle for Standards', Section 2 Committee 'Hot Topics' Discussion American Bar Association Section of Antitrust Law, Spring Meeting, Washington, D.C. 1 April 2004 (www.usdoj.gov/atr/public/speeches/203780.htm).

Monti, Giorgio, 'DG Competition Discussion Paper on the application of Article 82 of the Treaty to Exclusionary Abuses (December 2005), Comments from: Giorgio Monti, Lecturer in Law, Law Department, London School of Economics' (http://ec.europa.eu/comm/competition/antitrust/art82/065.pdf).

Muris, Timothy J., ' Competition and Intellectual Property Policy: The Way Ahead', American Bar Association Antitrust Section Fall Forum, Washington, DC, 15 November 2001 (http://www.ftc.gov/speeches/muris/intellectual.htm#N_2_).

Murphy, Anthony, 'Queen Anne and Anarchists: can Copyright survive in the Digital Age?', Speech at Oxford IP Research Centre Seminar, 26 February 2002 (http://www.oiprc.ox.ac.uk/EJWP0202.pdf).

Nalebuff, Barry, 'Bundling, Tying, and Portfolio Effects', DTI Economics Paper No 1, Part 1 – Conceptual Issues, February 2003.

Nalebuff, Barry and Majerus, David, 'Bundling, Tying, and Portfolio Effects', DTI Economics Paper No 1, Part 2 – Case Studies, February 2003.

OECD Policy Brief, 'What is Competition on the Merits?', June 2006, (www.oecd.org/dataoecd/10/27/37082099.pdf).

Padilla, A. Jorge, 'The Reform of Article 82: What We Agree, What We are Still Discussing and What will Have to be Discussed', Speech for LECG, Paris, 3 July 2007 (www.lecgcp.com/resources/documents/Library_300% 20-%20The%20reform%20of%20article%2082-2.pdf?pubtitle=LECG)

Pate, Hewitt R., 'The Common Law Approach and Improving Standards for Analysing Single Firm Conduct', Fordham International Antitrust

Conference, 23 October 2003 (www.usdoj.gov/atr/public/speeches/ 202724.htm).

Waters, Richard, 'EU launches new Microsoft antitrust probe', *Financial Times*, 14 January, 2009 (www.ft.com/cms/s/0/2ce90532-c2c1-11dc-b617-0000779fd2ac,_i_email=y.html))

Wilson, Bruce B., Deputy Assistant Attorney Gen., 'Patent and Know-How License Agreements: Field of Use, Territorial, Price and Quantity Restrictions' Remarks before the Fourth New England Antitrust Conference (6 November 1970).

Index